D0095257

MADELEINE CLARK WALLACE LIBRARY
WHEATON COLLEGE
NORTON, MASSACHUSETTS

Dripping Dry

Studies in Literature and Science
published in association with the
Society for Literature and Science

Titles in the series

Transgressive Readings: The Texts of Franz Kafka and Max Planck
by Valerie D. Greenberg

A Blessed Rage for Order: Deconstruction, Evolution, and Chaos
by Alexander J. Argyros

Of Two Minds: Hypertext Pedagogy and Poetics
by Michael Joyce

The Artificial Paradise: Science Fiction and American Reality
by Sharona Ben-Tov

Conversations on Science, Culture, and Time
by Michel Serres with Bruno Latour

Genesis
by Michel Serres

The Natural Contract
by Michel Serres

Dora Marsden and Early Modernism: Gender, Individualism, Science
by Bruce Clarke

The Meaning of Consciousness
by Andrew Lohrey

The Troubadour of Knowledge
by Michel Serres

Simplicity and Complexity: Pondering Literature, Science, and Painting
by Floyd Merrell

Othermindedness: The Emergence of Network Culture
by Michael Joyce

Embodying Technesis: *Technology Beyond Writing*
by Mark Hansen

Dripping Dry: Literature, Politics, and Water in the Desert Southwest
by David N. Cassuto

Dripping Dry

Literature, Politics, and Water in the Desert Southwest

David N. Cassuto

Ann Arbor
The University of Michigan Press

Published in the United States of America by
The University of Michigan Press
Manufactured in the United States of America
∞ Printed on acid-free paper

2004 2003 2002 2001 4 3 2 1

A CIP catalog record for this book is available from the British Library.

Library of Congress Cataloging-in-Publication Data

Cassuto, David N., 1963–
Dripping dry : literature, politics, and water in the desert Southwest / David N. Cassuto.
p. cm. — (Studies in literature and science)
Revision of the author's thesis.
Includes bibliographical references and index.
ISBN 0-472-09756-3 (alk. paper) — ISBN 0-472-06756-7 (pbk. : alk. paper)
1. American literature—Southwest, New—History and criticism. 2. Water supply—Political aspects—Southwest, New. 3. Politics and literature—Southwest, New. 4. Southwest, New—Politics and government. 5. Southwest, New—In literature. 6. Water supply—Southwest, New. 7. Water—Southwest, New. 8. Deserts in literature. 9. Water in literature. I. Title. II. Series.
PS277 .C37 2001
810.9'358—dc21 00-10715

Jesse Yates

Acknowledgments

This project began in 1992 at Indiana University as a dissertation idea. I walked into the office of a new faculty member I had never met and babbled something about my unfocused desire to write about water, the West, and literature. After we talked for a while, he leaned back in his chair and said: "It sounds like you're interested in water as a privileged signifier." "Yes," I thought. "Yes, that's it exactly." Cary Wolfe was that faculty member, and he proved equally instrumental at every phase of this project, lending erudition, support, and mentoring. Lee Sterrenburg directed the dissertation. Through his guidance, the project took form and came into its own as scholarship. Scott Sanders and James Justus also helped me refine my ideas into a workable thesis.

Many other people and entities bear prominent mention. I am grateful to the Indiana University School of Public and Environmental Affairs, which supported my graduate work though my degree was in English; to Dan Willard, my advisor at SPEA, who cheerfully believed in my work and in me before I gave him reason to do so; and to the National Endowment for the Humanities, which gave me a summer fellowship to work with N. Katherine Hayles—a privilege beyond reckoning. A number of my colleagues at the NEH seminar also offered invaluable guidance and criticism. I single out for special mention Stephanie Strickland and Joe Tabbi, whose insights and criticisms pushed me to say what I meant and to have what I meant to say mean something. I am also very grateful to Rob Dawson, whose photographs lend meaning to my words, no doubt because they say more about the region than words ever could. In addition, Julie Anderson spent many a precious summertime hour in the University of California, Berkeley, library sifting through photographs and helping me choose the images that enrich this book. Others deserving of special mention include J. D. Scrimgeour, Arthur Haubenstock, Carol Mitchell, and Ann Ronald, who gave generously of their time and insight. In addition, the chapters on Steinbeck and Abbey have appeared in different forms in *Papers on Langauge and Literature* and *ISLE,* respectively, and I am grateful to those journals for allowing their reproduction here.

I turn now to my loved ones. I come from a family of writers and editors. My brother Lenny and my parents Ike and Tobby Cassuto have been invaluable to this project. Each of them read, critiqued, and read again; never doubting but always questioning. My mother, a historian, also lent me her formidable research skills as well as her enthusiasm. There are no words for the kind of thankful I am.

Finally, my dear one, Elizabeth Downes, gets her own paragraph. It is not just for the clarity and incisiveness of her criticism and the generosity of spirit with which she offers it, but also for all the other things that, in attempting to describe, I could never do justice. Through two books now, she has been there for me in every possible way. During the home stretch of this one, she also carried our beautiful son, Jesse, to term.

I am many times blessed.

Contents

Introduction

Touch water and you touch everything.
—John Gunther

More than any other single characteristic, aridity defines the American West. It has shaped the land west of the hundredth meridian, the components of the social contract enacted among its inhabitants, and the compact between the inhabitants and the land.[1] Water's scarcity and its biologically critical function have also molded the regional literature. Using novels by Mary Austin (*The Ford,* 1917), John Steinbeck (*The Grapes of Wrath,* 1939), Edward Abbey (*The Monkey Wrench Gang,* 1975), and Barbara Kingsolver (*Animal Dreams,* 1990) this study aims to demonstrate how the myths pervading the regional literature interact with the myths that shape water policy, each helping to create the other, and both growing out of a limiting material condition.

The tensions and contradictions presented by these four novels underscore the compelling need for an ecocritique of the cultural symbiosis that exists between literature and politics. They further show that the need becomes especially urgent with regard to western water issues. *Dripping Dry* has two overarching aims: (1) to demonstrate the relationship between literature and politics with respect to the myth structure and policy choices involving water use in the West; and (2) to document the social consequences resulting from that relationship. More specifically, *Dripping Dry* focuses on the implications—both literary and ecological—of the Reclamation myth in the American West, as well as that of Restoration, its recent ideological rival.[2] A third myth, that of Sustainability, is explored in the concluding chapter.

Taken together, Austin, Steinbeck, Abbey, and Kingsolver traverse almost the entire twentieth century and therefore the entire Reclamation era (beginning in 1902, with the establishment of the Bureau of

Reclamation, and lasting till the present). Their works offer distinct versions of the conflict between the creations of an expanding, hydraulic society (i.e., dams, reservoirs, and other diversionary structures as well as energy and technology-intensive factory farm apparatuses) and the ecological realities of an arid region. In *The Ford,* Austin fictionalized the appropriation of the Owens River by Los Angeles, an event that came to symbolize the monopolistic water-grabbing that typified the Reclamation era. The novels of Steinbeck and Abbey explicitly participate in the dialogue over the efficacy of Reclamation policies. Kingsolver's integration of ecofeminism and Native American spirituality represents a compelling attempt to create a new vocabulary and system to supplant the extractive paradigm.

Examined collectively, the four works present a composite portrait of Reclamation. And Reclamation, I will argue, is one of the most important cultural and ecological phenomena in this nation's history. Born of the desire to graft the American yeoman ideal onto a sere western landscape, Reclamation eventually took on a life of its own. The supposed beneficiaries of massive water diversion projects became secondary to the projects themselves as a need for Reclamation for its own sake became ingrained in the national consciousness. Today, as the Reclamation era draws to an end, it leaves a cultural and ecological legacy that, for better or for worse, will remain with us for centuries.

Reclamation implies prior ownership (the root word is "*re*claim") and, in this context, also suggests a time when the land and its inhabitants enjoyed bountiful water resources rather than a water-scarce desert ecosystem. The underlying metaphor posits a contest between the forces of nature that illicitly appropriated the West's water, and human ingenuity, which seeks to reclaim it. John Wesley Powell, principal architect of the notion of Reclamation, spoke of "rescuing" and "redeeming" western rivers. Maintaining that "conquered rivers are better servants than wild clouds," he stressed the necessity of an ongoing war with the West's river systems ("Irrigable Lands" 767) and counseled that, for the arid lands, irrigation is "an absolute condition of continued prosperity" (776).

But Powell also advocated a pragmatic approach to irrigation that acknowledged the innate aridity of the western lands and the high cost of hydraulic projects. This cautious aspect of his philosophy was lost in the flood of enthusiasm for reclaiming the desert and making it bloom. By the turn of the century, the conservationist dogma propounded by the Roosevelt administration proclaimed that all the nation's assets (including the "reclaimed" West) should and could be put to human use. Americans fell back upon the rhetorical principles of their Puritan forebears, embracing Cotton Mather's doctrine, "That which is not useful, is vicious."

Mather's aphorism was reworked for the twentieth century by Gifford Pinchot, architect of the philosophy of conservation propounded by the Roosevelt administration. Pinchot advocated judicious stewardship of natural resources and attention to ecological realities, but only so that humans might better regulate nature and shape it to their aims. "The first duty of the human race is to control the earth it lives on," Pinchot declared (qtd. in Evernden 131). The central tenet of conservationism and its accompanying progressive ethic lay in the greatest good for the greatest number. Natural resources could and should maximize benefit to the populace.[3] Though mining and forestry also posed major concerns, water's use and control reigned as the West's most important issue. Powell's assessment of the region's hydrology, as well as his recommendations for the development of the land, forced decisions about the allocation and conservation of water resources.

Paramount among the desert's uses was, of course, enabling human habitability. Standards of comfort and habitability that depended on the climate and geography of the eastern United States were grafted on to the West. The idea that the arid region could be redesigned to meet eastern expectations of beauty and utility became an integral component of American myth.[4] It spurred the construction of many ecologically destructive dams and agricultural policies as its adherents strove to create a neo-Baconian Eden in a region with an average annual rainfall of less than twenty inches.[5]

By contrast, Restoration, a comparative newcomer to American mythology, looks to reconstruct ecosystems rather than subordinate them to human aims. As Carolyn Merchant describes it,

> Rather than taking nature apart and simplifying ecosystems . . . restorationists are actively putting it back together. Rather than analyzing nature for the sake of dominating and controlling it, restorationists are synthesizing it for the sake of living symbiotically within the whole. (*Revolutions* 268)

In the West, Restoration would involve returning the arid lands to their predam, preagribusiness condition—a presumed steady-state, ecological economy in which the region's scant hydraulic wealth is liberated.

Despite its ecofriendly veneer, Restoration, like Reclamation, raises serious land-use dilemmas. The realities of an urban nation, an expanding population, and an extractive economy combine with the ever-present western aridity to make Restoration seem quixotic and unproductive. Though small-scale restoration efforts—like reforesting a clear-cut or replanting a prairie—are often both successful and popular, it nevertheless remains true that, as a large-scale management strategy, Restoration has little chance of success. For example, Edward Abbey's ideal of independent agrarian communities subsisting in a restored desert landscape (discussed in chapter 4) appears unrealizable in the absence of a wholesale social and technological revolution. Some critics argue that this grail-oriented ecology impedes a more effective, practical approach.[6]

History has also shown that in regions where irrigation fuels agriculture, hydraulic empires are virtually inevitable. If so, Restoration seems even more untenable and Reclamation unavoidable. In many ways, Reclamation is merely a new word for an old phenomenon—an irrigation-based society run by the educationally and economically privileged. The ancient Nile civilizations, Chinese dynasties, Mayan civilizations, and others were dominated by elite groups capable of building and affording water delivery mechanisms.[7] The water they provided, like the capital they generated, flowed toward the rich and powerful. In the United States, reclamationists used the rhetoric of Jeffersonian yeomanism to disguise the gestation of a similar hydraulic regime.

While Abbey's bioregionalist, agrarian vision of Restoration may seem impractical, Restoration as a concept enjoys widening acceptance. This is understandable. A philosophy that hearkens to an identifiable past (the pre-Reclamation era) appears more inherently reasonable than one that conjures a fictional past (West as Eden) in order to justify an untenable future. Resource managers throughout the country have also begun to ally themselves with Restoration. Around the nation a number of rivers have made incremental headway in what will undoubtedly be a long and arduous struggle against an entrenched ideology and the results of its reign.[8]

Though encouraging, these gains cannot offset the continuing degradation of rivers and arid ecosystems in the name of Reclamation and western tradition. Recovering ecosystems are local phenomena, whereas the ideology of exploitation and hydrological mismanagement is national in scope. Any long-term solution to the West's ecological dilemma must be able to differentiate effectively between local and national problems and respond accordingly.

Furthermore, though the concept of Restoration is facially more reasonable than Reclamation, it lacks internal coherence. Restoring an ecosystem means returning it to a prior state, yet there is no one premodern ecological ideal. Ecosystems are in constant flux; their resiliency depends on their responsiveness to environmental disturbance. Therefore, there can be no one ideal state. Deprived of this imagined optimal prior circumstance, the concept of Restoration loses referentiality. Attempting to return an ecosystem to an idealized static condition has more to do with human cultural preferences than with the region's ecological health. William Jordan portrays restoration as means of achieving a harmonic relationship with a particular landscape (see Wilson 114–15). Unfortunately, as the Reclamation era has made clear, a perceived harmony with one's surroundings does not necessarily dovetail with the long-term health of the region.

The central issue for those seeking an ecologically acceptable alternative to Reclamation or Restoration must be the development of a rhetoric capable of expressing a nonexploitative but realistic relationship with nature. Given Reclamation's entrenchment and its historically powerful grip on the American Puritan imagination, the task seems especially difficult. Literature has played a key role in the creation and expression of the Reclamation era, and it must, I believe, play a similarly important role in the transition to a new era of Sustainability. *Dripping Dry* represents my attempt to aid in that transition.

Few would contest that our cultural conception of nature is a linguistic construct. By the same token, few would deny that nature exists independently of language. In discussing the relationship of facts to science, Alan Gross argues that "brute facts themselves mean nothing; only statements have meaning, and of the truth of those statements we must be persuaded" (4). The same could be said of our relationship with nature. Our surroundings exist, but they gain meaning only through their connection with us. And that connection is subject to constant reevaluation and scrutiny.

Yet nature becomes accessible only through socially and culturally specific languages, codes, and disciplines. Throughout this study, I make use of a double perspective, relying both on science-based ecological history and theory, and several schools of cultural criticism, including Marxism and ecofeminism.

In the concluding two chapters I also apply systems theory to the literature and ecology of the West, using its principles to illuminate problems born of the Reclamation era, and to explore possible solutions. In the final chapter, I argue that environmentalism, which has long lacked a satisfactory definition, needs to be redefined and resituated within the context of modern electronic culture. In many ways, environmentalism resembles a colossal hypertext—disparate causes linked by cybernetic bonds that gain meaning and influence through unity. The hypertext analogy allows me to negotiate the shifting terrain of a partially constructivist, partially objectivist vision of nature while enabling environmentalism to avoid solipsism and remain flexible, practical, and accessible.

Though literary and social theory figure prominently throughout this book, the final two chapters are more overtly theoretical than those preceding. That shift occurs for several reasons. The first four chapters trace the Reclamation era from its inception through the recent past (*The Monkey Wrench Gang*, published in 1975, frames chapter 4). That emphasis on the past enables me to draw upon a rich body of history and literature to illuminate and explain the ideas advanced in those chapters. The present and future—the focus of chapters 5 and 6—allow no such luxury.[9] Attempts to contextualize the present and shape the future require a certain level of abstraction—as would any endeavor treating matters that have not yet occurred with methods that do not yet exist.

Only the future can bring the present into relief, even as the future is itself a product of the present.

American civilization, particularly in the West, stands poised on the brink of an ecologically driven paradigm shift. Though the idea of an Eden in the desert and the rhetoric of Reclamation have lost much of their luster we have yet to craft sustainable alternatives. That task must fall to theory. Theory, however, must build on history and literature, and this book's structure represents my attempt to do just that. The ideas presented in the final two chapters are the culmination of the previous four and, I believe, reward the effort expended in their reading.

Envisioning a sustainable post-Reclamation reality in the West necessitates a radical reformulation of humanity's relationship with nature. Defining and creating nature has been and remains one of the central tasks of myth, and myth is of course a product of human thought and language. Both the Reclamation and Restoration myths attempt to build a cultural conception of nature sympathetic to their respective imperatives. Reclamation posits a national need to implement the American Dream. Like Restoration and multitudinous other land-use strategies, Reclamation offers a vision of nature that is both biological and historical. This is myth not in the sense of fable, but as an interpretive system that claims value neutrality while providing cultural legitimation for an ideological agenda. The myth of the West as Eden, better known as the myth of the garden, wherein there is bounteous water to meet every human need, exemplifies this phenomenon.

There are, Roland Barthes argues, three ways of reading myth (*Mythologies* 129). If I see a picture of the Hoover Dam and decide it is a symbol of American ingenuity and technological supremacy, then I am acting as a producer. If I believe the Hoover Dam does not just represent American technological supremacy, but *is* that supremacy, then I am a reader, accepting without question the union of symbol and signifier. Lastly, if I decipher the work of the producer by divining that a human agent decided what the dam symbolized in order to further a specific agenda, then I am a mythologist. Throughout the history and literature of the West, examples of the first two ways of reading myth abound. *Dripping Dry* implements the third mode, deciphering the myths of Reclamation and Restoration and then offering an alternative, Sustainability, in their stead.

This task looms difficult, both because of the size and complexity of the region and because nature continues to evolve and create itself, a process involving language, ideology, and the land. Analyzing their interaction requires an awareness of both the ideological forces at play and of the field on which they are playing. Neither semiotics nor science alone can solve our ecological crisis. Consequently, linguistic and science-based approaches function here as complements rather than opponents. I hope that a deeper acquaintance with the history of the West and its hydrological difficulties will better enable us to construct a new rhetoric of ecology and thereby locate and sustain sources of ecological renewal.

1

The Birth of the Hydraulic Conflict

> Where agriculture is dependent on an artificial supply of water, and there is more land than can be served by the water, values inhere in the water, not in the land; the land without the water is without value.
>
> —John Wesley Powell

In the eighteenth and nineteenth centuries, the American West changed slowly from a wilderness on the nation's edge to an integral component of the national consciousness. The journals and writings of explorers like Lewis and Clark and Zebulon Pike gave shape to a land previously unknown, while their descriptions led to policy decisions, migration, and further exploration.[1] Later, with the frontier closed and the nation committed to large-scale development of the West, policies arose that ignored ecological realities in favor of jingoistic Americana designed to propagate the notion of the West as Eden. Wishful thinking, coupled with a desire to remake the landscape in the image of human needs and wants, led to what Henry Nash Smith labeled the "Myth of the Garden." That which was not already Edenic would soon become so through human ingenuity and American perspicacity.

Imagining the land as virgin and Edenic ignored the geographical realities of a large indigenous population and a varied terrain and climate. Westward expansion, rather than puncturing these myths, fueled an extraordinary campaign to remake the wilderness in the image of that mythic landscape.[2] Ralph Waldo Emerson, echoing the national ethos of the mid–nineteenth century, counseled "action proportioned to nature." What nature does not yield freely, humanity should refashion to better suit human needs. This transformative relationship with nature harmonized, in Emerson's view, with nature's status as the ultimate commodity. Humans fulfill their destiny through working the land and forcing ever-greater harvests. In a section in *Nature* aptly entitled "Commodity," Emerson maintains that nature has no greater purpose than to serve "Man." And Man has no greater purpose than to work the land and take his place in the productive cycle: "A man is fed not that he may be fed but that he may work" (*Selections* 26). Emerson's views were and continue to be widely shared. The policies born of such views have led to ecosystemic catastrophe, the full implications of which are yet to be felt. The bulk of the damage occurred during the twentieth century as technological innovations permitted greater and greater short-term dominance over the land.

The Roots of Reclamation

The Bureau of Reclamation was born in 1902 to bring water wherever it was required. The impetus for reclamation lay in its ties to yeomanism, a concept whose roots lay in Thomas Jefferson's vision of the United States as an agrarian democracy. Jefferson's dream of a nation of independent farmers living off of the land captured the popular imagination. With western expansion, that dream grew more enticing still. However, implementing it in the arid West proved impossible without technological intervention. Providing water to sustain both the myth of the garden and the settlers who came West believing in that myth became the task of the Bureau of Reclamation. After a shaky beginning, the bureau quickly grew powerful, and it, possibly more than any other human contrivance, shaped the geography of the West. Consequently, the years from 1902 through the present (although the bureau's power has waned considerably in recent years) have come to be known as the Reclamation era.

The Reclamation era refers to a time of unprecedented development fueled by previously unthinkable hydraulic projects. Reclamation projects grew steadily more ambitious as the century progressed, hitting full stride between the 1930s and 1960s, but maintaining a frenetic pace into the 1980s.[3] Marc Reisner has termed the halcyon days of reclamation the "Go-Go Years." During those years, water projects sprouted like mushrooms throughout the arid lands.

Massive dams and diversion projects brought water to areas once considered irredeemably dry. Hoover, Shasta, Bonneville, and Glen Canyon Dams, among others, stand as monuments both to hydraulic engineering and to the Bureau of Reclamation's deep pockets. Enormous amounts of water suddenly became available for use, leading to the misconception that there was abundant water. The newfound ability to extract water from previously inaccessible or impractical sources was misrecognized as the ability to create a water supply.

Actually, the surge in development resulted in *less* water because the reservoirs in the desert exposed thousands of acres of increased surface area to the sun's rays, dramatically increasing evaporation. Increased agriculture also caused large amounts of water to percolate into the ground rather than return to the rivers. Nevertheless, the myth of the garden thrived, unfettered by empiricism. As Mark Seltzer observes, "Nothing typifies the American sense of identity more than the love of nature (nature's nation) except perhaps the love of technology (made in America)" (3).

The Bureau of Reclamation was a predictable outgrowth of a nation "accustomed to plenty and impatient with restrictions." Americans chose to deny aridity's existence for a while and then, when that was no longer possible, to "engineer it out of existence or to adapt to it" (Stegner, *Bluebird* 75). The latter choice was anathema, while the former strategy has ruled the West for much of its history.

In the mid–nineteenth century, advocates of western settlement insisted that "rain would follow the plow." According to this theory, Americans needed but to move west and till the land; rain would fall as a direct consequence of their labors.[4] By the 1880s, this notion had fallen into disfavor, but not before many settlers, lured by the vision of hydrological abundance, homesteaded the arid region.

The most ardent proponent of the rain-follows-the-plow credo was William Gilpin, first territorial governor of Colorado (1861–62) and indefatigable apologist for the Plains. He viewed the arid lands "through a blaze of mystical fervor," envisioning the Southwest as a bountiful paradise that awaited only the human touch to yield unimaginable bounty (Stegner, *Meridian* 2). Gilpin trumpeted manifest destiny and the boundless munificence of the West in books with grandiose titles including *The Mission of the North American People* and *The Continental Railway, Compacting and Fusing Together All the World's Continents.*

Gilpin words were buttressed by his considerable personal credibility. He had fought with the Missouri Volunteers against the Mexicans in 1846 and later joined expeditions against the Comanche and Pawnee tribes. At the time of his appointment as governor of Colorado, Gilpin had been serving as a volunteer bodyguard to President Lincoln. His reputation as an intrepid western explorer and soldier coupled with his political connections lent great weight to his words in both government and civilian circles.

In Gilpin's view, the West possessed a virtually limitless ability to sustain ever-growing numbers of immigrants in affluence and comfort. Since one needed only to turn over the soil in a western homestead for the heavens to release the requisite moisture for agriculture, the Mississippi River basin alone could, according to Gilpin, house 1,310,000,000 people. The Plains loomed even more sylvan and grandiose: "The PLAINS are not *deserts,* but the OPPOSITE," he proclaimed. They formed "the cardinal basis of [a] future empire" (*Mission* 66). In Gilpin's eyes, the western United States contained a reservoir of resources, including land, water, and precious metals (especially gold) that would readily yield to the energy and ingenuity of American pioneers. Gilpin's panegyrics of the West and his pseudoscientific explanations for the region's alleged munificence were not unique.[5] His views came to prominence in the 1860s, but they built on an already well established tradition of western self-promotion.

Thomas Hart Benton, senator from Missouri, had originally believed that the Plains could best serve the nation as a conduit to the West Coast and then to India. Later, as he became convinced of the ability of the railroads to open the region to commerce and settlement, Benton began trumpeting the region as a bucolic paradise. Speaking in Boston in 1854, Benton described Kansas as "rich like Egypt and tempting as Egypt would be if raised above the slimy flood, waved into gentle undulations, variegated with groves and meadows [and] sprinkled with streams" ("Discourse" 4).

Benton's rhetoric, coming as it did from a Southerner, was matched exclamation for exclamation by Northern free-staters like Charles Boynton and T. B. Mason. They proclaimed that the Kansas landscape contained "many scenes that can scarcely be *remembered* without tears. The soul melts in the presence of the wonderful workmanship of God" (qtd. in Emmons 14). Such paeans to the majesty and fertility of the region reflect more than simple optimism and a willingness to overlook obvious geographical traits. At the time, the North and South were locked in a high-stakes battle for social and political dominance of the Plains.

The Kansas-Nebraska Act of 1854 allowed residents of those territories to choose their status as either free or slave states. The balance of power in the nation, held uneasily in place by the Missouri Compromise (1820–21), stood ready to shift. In the North, the prospect of slavery in the Plains meant the imminent demise of the yeoman farmer. Slaveholders would occupy vast tracts and work them cheaply with slave labor. Independent yeomen would be excluded from the best land, their voices silenced by the economic might of the Southerners. Such a fate would be "a gross violation of a sacred pledge, . . . an atrocious plot" aimed at the heart of the yeoman ideal (Charles Sumner and Salmon Chase, qtd. in Emmons 12).

The enactment of the Homestead Act of 1862, which deeded 160 acres to anyone willing to settle and work the land in the West, marked an enormous victory for Lincoln's Republican party over the proslavery forces of the South. The Southern plantation system required large amounts of land, often exceeding 1,000 acres, to function profitably. Strict acreage limits on homestead properties meant that slaveholders would have great difficulty gaining a foothold in the Plains. That handicap effectively eliminated the South from further participation in American westward expansion.

The next challenge facing the Republicans lay in finding people willing to move west. Likely candidates included poor people in the eastern United States and in Europe—people who would be enticed by the promise of free land and a new start. Principal responsibility for the campaign to entice immigrants to the West fell to the railroads. While westward migration served the interests of a young nation with an expanding population and a vision of manifest destiny, it also benefited the emergent transcontinental railroad. By the mid–nineteenth century, these congruent interests formed the basis of a potent alliance.

The federal government ceded vast tracts of land to the railroads to open up the West, facilitate trade, and generally strengthen the nation. In return, the railroads marketed the land and promoted settlement. Railroad officials took to their task with an enthusiasm that bespoke the enormous profits they stood to reap from increased western settlement. Promoters (called "boomers") circulated literature proclaiming the Plains an agricultural mecca. One pamphlet even insisted that "mud in the usual sense . . . is almost wholly unknown in Nebraska" (qtd. in Baltensperger 58).[6]

The boomers' unbridled enthusiasm and their willingness to stretch the truth are reminiscent of the campaign by the Associated Farmers in the 1920s and 1930s to lure Dust Bowl refugees to California. In each instance, propagandists used the symbol of the yeoman farm and the promise of an agricultural paradise to lure people to a region that little resembled their descriptions. In the Plains, the railroads and the federal government stood to gain from increased immigration to the region but cared little what befell the immigrants once they arrived. In California, corporate growers already controlled the land and water and were merely looking for a new workforce to exploit. The common denominator in the two examples is a widespread reliance on the myth of an Edenic garden on the nation's frontier coupled with an unwavering faith in manifest destiny.

The notion of a camouflaged Eden with no historical link to European traditions permitted Americans to reincarnate themselves in a "virgin land."[7] The rugged new continent, rather than representing a continuation of European cultural hegemony, offered a singular destiny for those brave enough to seize it. Without an acknowledged history, America offered a new beginning wherein land and settler could merge into a single entity and recover—through diligence, husbandry, and mettle—the lost paradise of Eden.

Myra Jehlen argues that this vision embodied the American tendency to merge selfhood into a collective national ideal while preserving a uniqueness defined by one's own relationship to the land. Settlers inscribed their own identity on to the tabula rasa of the new continent, using its wildness as a means to self-discovery. Only through subduing the wilderness, however, could the new nation take shape and thereby confer a national identity on to its citizens. The result was a land-based Hegelian opposition; civilization and wilderness formed one thesis and antithesis, and humans and nature were another. Each half of the opposition provided the other's "cathartic," and provided the means to "the emergence of the single and unchanging truth" (82).

For nineteenth-century settlers in the West, the opposition lay in the juxtaposition of aridity with the Jeffersonian yeoman ideal. Their synthesis created the "truth" of the yeoman plains farmer. Rather than abandoning the myth of the garden in the face of looming ecological realities, Americans chose to redesign the land and carve a garden out of the desert.

In the twentieth century, the vision of the family farm transplanted to the arid lands served as the Bureau of Reclamation's raison d'être and a convenient propaganda tool for corporate capitalism, as agribusiness interests sought and gained control over much of the region's land and water. The resulting tension between societal myth and social and ecological realities infuses the regional literature. Understanding the complex interweaving of policy and fiction surrounding water and its appropriation requires that we first survey the historical underpinnings and embedded myths of the Reclamation era.

Powell and His Legacy

John Wesley Powell (1834–1902), a major in the Union army who lost an arm at Shiloh, went on to become one of the most important figures in the history of western exploration. Self-educated as both a geologist and anthropologist, Powell led a survey team through the previously unexplored Colorado River canyons in 1869. He and his eight companions were the first white people and perhaps the first of any race to successfully traverse the entire treacherous length of the Grand Canyon. Their success carried a heavy cost. Three members of the expedition died.

They also lost three of their four boats and most of their supplies. Their journey took them through previously uncharted terrain, and their courage in the face of the unknown and amid bone-crushing rapids turned them into national heroes. Powell also viewed the journey as a scientific expedition; he took measurements as carefully as he could and sought to tie his observations to a larger geological survey of the region.

During the expeditions, Powell and his crew gave names to the formations and canyons they encountered.[8] Their sense of wonder and reverence for the terrain—captured in names like Marble, Glen, and Music Canyons, Bright Angel Falls and Rainbow Bridge—gave a shape to this canyon country that most of the East Coast–based population had never seen. The fame Powell garnered from his journey, in concert with his ferocious work ethic, helped him rise through the ranks of government to wield enormous power in western land and water issues. As head of what later became the federal Irrigation Survey, Powell's efforts led to the mapping of much of the West.

After an initial rise to prominence, though, Powell's views fell into disrepute in Washington, and his reputation suffered. His fall from grace stemmed primarily from differences with powerful western senators over matters pertaining to land and water appropriations.[9] However, his reputation was rehabilitated in the early twentieth century, as conservationists selectively adopted his ideas. In a sense, the entire Reclamation era owes itself to Powell, although he probably would not have wished credit for much of it.

Powell's 1878 *Report on the Lands of the Arid Region of the United States* was a revolutionary document. Ignored at first, and then selectively implemented, the report may well be the most important document in the history of American land use. It offers the first comprehensive management strategy that acknowledges the region's ecological parameters. One of Powell's first tasks in the treatise was to debunk the notion that rain follows the plow, a theory for which he had little patience:

> [T]he operations of man on the surface of the earth are so trivial that the conditions which they produce are of minute effect, and in presence of the grand effects of nature escape discernment. Thus, the alleged causes for the increase of rainfall fail. (*Report* 91)

He further rejected the quixotic faith in technology that would supposedly create a new Eden in the desert. Instead, he proposed a system whereby the federal government would regulate the distribution of land and the construction of water projects so as to allow individual settlers to profit from the national trust. Noting that "all the great values of this territory have ultimately to be measured . . . in acre feet"[10] (qtd. in Stegner, *Meridian* 315), Powell urged that the current 160-acre homestead allotment be abandoned and that tracts be instead allocated according to their water rights.

Irrigated farms with guaranteed sources of water could produce crops even more reliably than farms in the humid regions that depended on rainfall. An arid-lands farmer with access to a reliable water source could therefore live comfortably on fewer acres. Powell proposed an 80-acre limit for irrigated quadrants. Nonirrigated tracts, depending on their location, could graze cattle or other livestock, but would need considerably more land to do so. He consequently suggested that nonirrigated pasturage farms contain 2,560 acres, that rangeland not be fenced so that livestock could graze communally, and that the vast western lands be treated as a sort of a commons (*Report* 22–23). The cumulative effect of these policies would have been dramatic. Not only would they have drastically curtailed land monopolies, but the precedent of adapting human use to ecosystemic parameters would have become a matter of law.

Perhaps more importantly, *Lands of the Arid Region* advances the idea of government control over water distribution. Yeoman farmers could not possibly finance the massive diversion projects required to bring water to many sites. Left to their own devices, they would have to either sell out or purchase water from those wealthy enough to construct delivery devices. Water monopolies would inevitably result from such an arrangement, to the impoverishment of small farmers.

> If the water rights fell into the hands of irrigating companies and the lands into the hands of individual farmers, the farmers will then be dependent upon the stock companies, and eventually the monopoly of water rights will be an intolerable burden to the people. (41)

Powell displayed an uncanny prescience in his admonition against the monopolization of water rights and in his vision of its effects on western lands and people. He anticipated the enthusiasm for unchecked development and took the unpopular stance of advocating prudence and regulation:

> [I]f in the eagerness for present development, a land and water system shall grow up in which the practical control of agriculture shall fall into the hands of water companies, evils will result therefrom that generations may not be able to correct, and the very men who are now lauded as benefactors to the country will, in the ungovernable reaction which is sure to come, be denounced as oppressors of the people. (41)

Powell was, in effect, proposing a hydraulic welfare state in the West, an idea wholly counter to the popular notions of rugged individualism and hydroabundance, as well as the designs of the hydraulic empire builders whose power and interests were already well entrenched.[11]

Virtually all of Powell's proposals died in Congress, victims of "the myth-bound West which insisted on running into the future like a streetcar on a gravel road" (Stegner, *Meridian* 338). Several decades later, at the end of the nineteenth century, a prolonged drought belatedly convinced legislators to revisit many of Powell's recommendations. Powell died in 1902, the same year that the passage of the Newlands Act paid posthumous tribute to his notion of a government-controlled hydraulic infrastructure by creating the Bureau of Reclamation. Powell, legislators admitted, was right; without government intervention, water monopolies would control what little water became available while other, larger projects would never come to pass due to lack of funding.

With the creation of reclamation districts and federal control over water distribution, legislators believed the aridity dilemma solved. They ignored Powell's warnings that the land-parceling system also badly needed reform. Without a revised system of land management, however, government subsidization of water projects during the first quarter of the twentieth century simply helped entrench corporate control of the region's water. Small farmers still could not compete with large landowners when both benefited from water subsidies. Factory farming gained strength in the West, and a mutually profitable relationship arose between the bureau and wealthy agribusinesses. The repercussions—

both human and ecological—of this relationship become glaringly apparent in the 1930s when the black blizzards and the Dust Bowl hit the Plains.

The Bureau of Reclamation justified its existence and its unprecedented power to shape the landscape by building dams and canals that generated hydroelectricity, provided nominal aid to small farmers, and watered the fields of the agribusiness giants. The incredible flurry of dam building that took place between the 1930s and the early 1980s left the country with over one hundred thousand dams. Most of the dams were built either by the bureau or by the Army Corps of Engineers. During this era, the Army Corps of Engineers and the Bureau of Reclamation pursued an expensive rivalry, attempting to one-up each other by building bigger and more expensive dams.

Whereas the Army Corps of Engineers' ostensible justification lay in its stated mission of flood control, the bureau relied on the yeoman ideal and the myth of the garden to justify its actions. The bureau's dams purported to make the desert bloom, thus enabling rugged individualists to wrest their livings from the land. That scenario was, of course, woefully inaccurate. Corporations, not individual farmers, were the primary beneficiaries of the bureau's largesse, and, despite the bureau's herculean efforts at reclamation, the total irrigated area in the West remains to this day smaller than the state of Ohio.[12]

In the 1950s, even as dam construction in the United States was reaching its zenith, a new doctrine of environmentalism, specifically the concept of Restoration, began to compete with the conservation rhetoric of the Bureau of Reclamation. In the postatomic era, faith in technology as the panacea for all of humanity's problems began to ebb. Though it continued building dams and diversion projects at a ferocious pace, the bureau also suffered several high-profile setbacks. Proposed dams in Dinosaur National Monument and the Grand Canyon met with ferocious public resistance and were shelved. Critics of reclamation projects cited human comfort as but one of many considerations in the land-use debate. The value of rivers and watersheds as something more than fodder for fields, faucets, and turbines demanded consideration. In addition, the fiction of limitless water began to fray.

The problems arose because policymakers had ignored a crucial component of Powell's management philosophy—the need to work within the region's hydrological boundaries. The Bureau of Reclamation and its allies maintained that technological advances made natural characteristics all but irrelevant to policy decisions. Powell, they claimed, would have agreed. Much of the rampant development wrought throughout the West during the twentieth century supposedly bore Powell's imprimatur. On the other side of the issue have been those who opposed the projects, also claiming Powell's posthumous approbation by pointing to his pragmatism and strident opposition to water and land monopolies.

The central symbol of the contemporary tug-of-war over Powell's postobituary blessing is Lake Powell, the massive reservoir created by the damming of the Colorado at Glen Canyon in 1963. Powell had given Glen Canyon its name during his explorations of the Colorado and had rhapsodized about the region's incredible beauty. The bureau then gave Powell's name to the reservoir that flooded the canyon, contending that Powell would have been pleased because the dam supplied hydroelectric power to a large area while also providing irrigation benefits. In addition, the dam made the unflooded parts of the canyon more accessible, thereby serving the greatest number.

Critics of the dam argue that Powell would have blanched at the tremendous waste of resources that the dam incurred as well as at the destruction of what many believe had been the most beautiful canyon system in the country. Evaporation losses from the reservoir amount to millions of acre-feet per year; irrigation benefits from the dam are comparatively minimal, and the precious silt carried by the Colorado—silt that provides fertility as well as a buttress against downstream bank erosion—backs up behind the dam, where it serves no use.

In the aggregate, Powell's writings and life's work suggest that he would have likely deemed the Glen Canyon Dam a waste of precious resources as well as a violation of the Bureau of Reclamation's stated mission. Furthermore, Powell understood better than many of his more educated descendants the importance of thinking in terms of entire watersheds rather than small sections of rivers. Glen Canyon Dam causes significant harm to downstream ecosystems, while providing incommensurate gain to its beneficiaries. For that reason also, Powell would not have endorsed it. Still, the dam's very existence, as well as the increased settlement of the Four Corners region, endorses the populist agenda that Powell championed. It would seem that Powell's vision, like the West's geography, inspires great passion while remaining eminently mutable.

Western Water Law

Over the course of its settlement, the West has variously been labeled the Great American Desert, the Promised Land, a worthless terrain, and the garden of the world. Ultimately, one's perception of the region hinges on the availability of water, which determines whether one can survive and how well. Water's primacy as both commodity and cultural sign in the Southwest arose on this foundation of scarcity and utility.[13] Its biotically privileged place in the culture predates its commodification, but the two traits have long been inseparable in the history and mythology of the American West.

The power to control water carries with it the power to control life. And by "life," I refer not just to existence, but also to *quality* of life. Many of the features that define the "good life" require water and revolve around its distribution and control. In addition, since water is necessary to human biological function, in an arid region a dominant state apparatus would need to expend relatively little effort to transform water into a commodity whose scarcity privileged the water as well as its controllers. That scarcity would boost water's exchange value, leading to intensified symbolic value. Symbolic value imparts power and wealth and thereby privileges its possessor. In this sense, water becomes not just a measure of economic value, but a culturally privileged signifier as well.

Water's emergent status as commodity and signifier of wealth and expanding capabilities forced an important shift in garden mythology. The Newlands Act amounted to an implicit acknowledgment of water's scarcity as well as the limited productive capabilities of the land. Such an admission also involved conceding the limitations of the nation and its people: a prospect that remained anathema to a culture steeped in the dominant myths. Jehlen notes that "the conviction that farming brought reason and nature together (since man and nature had the same reasons) inspired cultivation . . . but made it particularly difficult, in fact, contradictory to contemplate basic changes in agrarian policy" (73). Instead of abandoning the American Dream, the dream itself shifted focus. The myth of the garden remained intact, but its ideal evolved from a Xanadu to a neo-Baconian Atlantis that no longer awaited manna from heaven, but wrested it instead from the grips of Nature.

The evolution of western water law offers an interesting perspective on the West's adaptation to water's scarcity. Water law in the eastern United States is based on riparian rights and evolved from the common law of England. In general, riparian water rights allow anyone owning property bordering a watercourse to make reasonable use of its water as long as that use does not interfere with the reasonable uses of other riparian landowners. In the East, where water is abundant, the reasonable-use doctrine emerged as workable and fair.[14] Twenty-nine states continue to adhere to riparianism today.

In the West, however, riparianism proved untenable. During the time of rapid westward expansion in the nineteenth century, demand for water, especially for mining, far surpassed supply. Miners faced a dual dilemma. First, the riparian system restricted water use to those who owned land bordering streams, while the demand for water often lay elsewhere. Second, most of the land in the West was owned by the United States. Miners, though squatting on the land with the tacit approval of the federal government, were technically still trespassers and could not acquire riparian rights.

In the face of these restrictions, rules for allocating water developed in the mining camps and gradually spread throughout the West. These rules were similar to those for establishing and protecting mining claims. The first user of water from a specific source held a right to that water that was superior to all subsequent claims. As long as that claimant made "beneficial use" of the diverted water within a reasonable time, s/he held the water right in perpetuity.[15] The federal government recognized the validity of this system of allocation in early mining acts and later in the Desert Lands Act of 1877. Thus developed the doctrine of *prior appropriation* and its credo of "First in time, first in right." Prior appropriation dominates western water law to this day.[16]

The Jeffersonian ideal could not survive unscathed in a region of intense competition for water, a region growing ever more beholden to agrocapitalism. Farmers had to seek profit rather than subsistence, a process that in turn impelled them to place more and more land under plow. They also needed to irrigate. Except for those lucky few whose acreage bordered a water source, irrigation required diversionary structures and a water right. Neither came cheap. The Homestead Act of 1862, instead of providing land for small farmers and encouraging the yeoman ideal, facilitated speculation, corporate land grabs, and water monopolies.

The Homestead Act had originally provided for 160 acres of land for all eligible applicants. In 1909, the realities of "dry farming" impelled Congress to pass the Enlarged Homestead Act, doubling the amount of land available to settlers. Neither act accomplished its goal of making small parcels available to underfinanced settlers. Nor was either parcel big enough, without an accompanying water right, to allow settlers to compete with large, well-financed agribusiness interests, many of whom had entrenched themselves in the West with the unintended aid of the earlier Homestead Act, as well as the Desert Lands Act.[17]

Technological advances in farming techniques meant greater yields, which translated to greater supply and lower-priced goods. The arrival of the tractor and the one-way disk plow in the early twentieth century ushered in an era of unprecedented agricultural productivity. In 1830, it took fifty-eight person-hours of work to bring an acre of wheat to the granary. By 1930, in parts of the Great Plains, that number had been reduced to three (Worster, *Dust Bowl* 90–91). Even as increased productivity meant greater supply and falling profit margins, the price of the machines that enabled that productivity continued to rise. Farmers found themselves putting more and more land under plow just to maintain their current income level. Many fell into debt, which in turn forced them to plant still more, ensnaring them in an increasing cycle of debt and ecological degradation that eventually led to the Dust Bowl of the 1930s. Small farmers could not compete; their debt burden soared, and more land and water fell into the hands of corporate interests, where it has since remained.

Four Visions of Water and the West

The chapters that follow divide the Reclamation era into four segments and examine each through the lens of a particular work of fiction. The four authors treated—Austin, Steinbeck, Abbey, and Kingsolver—all concern themselves deeply with ecological and human issues regarding land and water in the West.

Mary Austin found herself caught between her nascent environmentalism and love of the desert (visible throughout her canon and particularly in such works as *Land of Little Rain*), and her allegiance to conservationist dogma. *The Ford* offers an excellent look at the tensions between ecological awareness and an extractive economy. It is a roman à clef, a fictionalized version of Los Angeles's illicit appropriation of the Owens River from the Owens Valley in northeastern California, that portrays the conflict between urban and rural interpretations of conservation. Underlying the conflict is the ideological gulf separating proponents of rural agriculture from those who favored an increasingly urban western landscape.

Austin lived in the Owens Valley at the time of the controversy and decried the "theft" of the water and the disregard for the land's integrity that such a maneuver entailed. Yet she fully supported the valley inhabitants' desire to dam the river for agricultural purposes. The conflict, as Austin saw it, lay in which use of the land and water was more ethical (within the ethical system that created the myth of the garden), not whether the ecosystem ought to be left intact. The collision between Austin's protoenvironmentalism and her allegiance to the Pinchot-derived progressive agenda demonstrates the lack of coherent alternatives within an extractive relationship with nature. Her dilemma effectively exposes the inchoate glimmerings of environmentalism amid the emerging juggernaut of the Reclamation era. Those who might have resisted the Bureau of Reclamation's rise to power lacked the vocabulary with which to question its conservation rhetoric.[18]

Eventually, the bureau came to dictate who would share in the hydraulic wealth that it alone was capable of distributing. Water's utility (use value) and its status as a commodity (exchange value) became ineluctably linked as the economy of the West swelled through its newfound water wealth. When its use and exchange value merged, water truly became the region's privileged sign, thereby allowing the bureau to secure its dominance in the West.

The events depicted in *The Ford,* both historical and fictional, serve as an excellent foreground to the ensuing discussion of *The Grapes of Wrath.* By the 1930s, the Bureau of Reclamation enjoyed virtually unquestioned authority and an unlimited budget for water projects in the West. That the primary benefactors of the bureau's engineering feats were corporate interests caused little stir in Washington or in state governments. Nevertheless, the impact of the boom in settlement and cultivation in the arid regions and the trend toward ecologically harmful factory-farming techniques soon made itself felt in the form of the Dust Bowl and in the human catastrophe of the subsequent mass migration to California. *The Grapes of Wrath* uniquely captures the magnitude and scope of this tragedy.

Throughout the novel, water serves as a privileged sign whose power is directly linked to its absence. Water is visible (or rather, invisible) in the drought that led to the Dust Bowl, the Joad family's cross-country pilgrimage, and in the family's frantic laboring during California's dry season to escape the starvation and poverty that would accompany the coming rains.

Unlike *The Ford* (1917), which had little measurable impact on the controversy it depicted, *The Grapes of Wrath* (1939) became a powerful force for social change. Much as Powell's *Explorations of the Colorado River* (1875) offered Americans their first glimpse of the canyonlands, Steinbeck's novel provided many with their first glimpse of the scope of the Dust Bowl and of the mistreatment of migrant workers.

Primarily as a result of the attention focused on the situation by *The Grapes of Wrath,* as well as several other contemporary works, officials at both the state and federal levels took their first tentative steps toward land-use reform and migrant worker protection. These steps marked the first significant national stirrings of discontent about the water and land-use policies emerging from the Reclamation era.

That discontent later grows into full-scale rebellion, as evidenced by Edward Abbey's advocacy of "ecotage," or environmental sabotage, to bring the Reclamation era to an end. *The Monkey Wrench Gang* (1975) not only examines the ramifications of contemporary western water policy, but also champions its demise. To Abbey, the Glen Canyon Dam exemplifies the cycle of what I call *faux potlatch* (see chap. 4), carried on by development interests with the aid and complicity of the government. Rather than reenacting the original Native American potlatch—which involves the large-scale accumulation of resources in order that they might be ritually expended—agribusinesses enjoy a corrupted version of the potlatch that is endemic to capitalism. Southwestern faux potlatchers amass huge quantities of the region's most precious resource (water) and then waste it in a manner calculated to incur maximum personal gain. This cycle is visible in the widespread cultivation of water-intensive crops (alfalfa, cotton, rice) and cattle raising in the West accomplished with subsidized water from oversubscribed rivers.

The agribusiness consortia learned that the road to greater government subsidy lay in expending as much water as possible through maximizing their cultivation of ecologically untenable crops and livestock. Western irrigation and reclamation projects spiraled out of control, freed of any obligation to justify themselves or their cost. The irrigation program envisioned by Powell had morphed into a hydraulic juggernaut that was growing steadily more powerful and impervious to criticism.

Abbey attacks the ideology and mechanisms of the Reclamation era, offering rough-hewn solutions born of anarchist leanings. In his view, the gang's mission to destroy the Glen Canyon Dam is moral within an ethical system defined not by Reclamation ideology, but by allegiance to the principles of the land ethic: "A thing is right when it tends to preserve the integrity, stability, and beauty of the biotic community. It is wrong when it tends otherwise" (Leopold 262).[19] Out of this ethos sprang ecotage—acts of sabotage committed to prevent environmental destruction. Earth-First! (whose symbol is the monkey wrench), Greenpeace Action, and other radical environmental groups owe much to Abbey's popularizing of the notion of ecodefense.

The Monkey Wrench Gang's assault on the ideology and mechanisms of Reclamation invites a discussion of alternatives to that ideology and the feasibility of their implementation. Barbara Kingsolver's *Animal Dreams* (1990) attempts to construct just such a scenario, and the novel's successes and stumbles provide an intriguing window onto an inarticulable future. Contemporary definitions of environment derive from the communicative mechanism of a system dedicated to extraction. *Animal Dreams* presents an attempt to supplant the dominant system (and discourse) with one privileging a steady-state society. Kingsolver advocates a worldview based on organicist ecofeminism and Native American ecospirituality instead of the mechanistic, patriarchal science that has caused what Carolyn Merchant calls the "death of nature."

Systems theory posits that environmental problems do not exist until they generate communication within the system. Environmental disturbances (resonance) are relevant in direct proportion to the systemic reaction (communication) they engender. In Niklas Luhmann's view, "Fish or humans may die because swimming in the seas and rivers has become unhealthy," but so long as this "is not the subject of communication it has no special effect" (*Communication* 29). In other words, until it is articulated, a disturbance, no matter how ecologically significant, will not affect the system. It follows, then, that if communication about a disturbance can be suppressed (i.e., through lack of language to describe it), the system's functioning will continue unimpaired.

The residents of Kingsolver's fictional town of Grace, Arizona, already have little capacity for resonance because their status as a Hispanic/Native American traditional community with a strong female presence leaves them with little social or economic clout. Interestingly, while the novel's protagonist spearheads the grassroots resistance that succeeds in stopping the destruction of the all-important river running through Grace, the town's environmental victory gets subordinated to her romantic reawakening. The novel's denouement downplays the ecosystem and instead stresses the successful reproduction of community.

At novel's end we are left with a term, *environment,* without a consensus definition. Uncertainty about the meaning of environment throws notions of ecology into flux and opens the way for a new rhetoric of environmentalism more sympathetic to alternative ideological and economic systems. Creating that new lexicon is crucial; without it, environmentalists can react to new antienvironmental initiatives but cannot propose workable alternatives to the status quo. However, the status quo is not acceptable; it privileges extraction and profit over ecological and social health. Environmentalists thus find themselves in the unenviable position of having to fight to maintain an unacceptable state of affairs because they cannot articulate any satisfactory alternatives.

Dripping Dry's concluding chapter discusses the long-term implications of this linguistic gulf. I believe our current ecological dilemma offers the potential for an exciting synthesis of science-based approaches to ecological management and language-based, theoretical methods. Neither strategy functions alone; each helps shape the other. Together they create the myth system that defines social and ecological realities.

Tellingly, the three key words informing this study and pervading the western hydrological morass all lack coherent definitions. *Reclamation,* as noted earlier, implies an imaginary previous time of hydraulic wealth and human dominion. *Restoration* is equally nonreferential. Making things how they were is neither possible nor productive. The arid lands evolved from a pre-Pleistocene sea and that process of gradual change will never cease. Ecosystems are always in flux; therein lies their ability to adapt and withstand disturbance. Indeed, the stability of the desert ecosystem—as with any ecosystem—lies in its evolution. Finally, *environment,* like *nature,* remains one of the most elastic words in the language. Its definition reflects its usage, which in turn reflects the needs and ideology of the user.

Defining these terms will only become more problematic as the ordering systems we use to shape reality gain complexity. Our society's escalating reliance on cybernetics and information theory make attempts to draw boundaries around the self or language increasingly quixotic. Using an approach that might be called historicist constructivism, but which I prefer to call *linked causes,* I suggest that we abandon the losing battle against complexity and instead seek to make the language and practice of ecosystem management reflect the infinite possibilities of existence.

2

River Wars: Los Angeles, the Owens Valley, and Mary Austin's *The Ford*

> But we have yet to mention the chief blessing of aridity. This is the fact that it compels the use of irrigation. And irrigation is a miracle.
>
> —William Smythe, *The Conquest of Arid America*

Lesta V. Parker, a longtime resident of the Owens Valley, wrote the following letter to President Roosevelt around 1905 (I quote it with spelling and punctuation intact).

> Pres. Theodore Roosevelt
> Washington D.C.
>
> Dear Friend:
>
> Look on your map of California, along the Eastern boundary south of Lake Tahoe and you will find a county named "Inyo." Running onto this county from Nevada through a small corner of Mono Co. you will see the Carson and Colorado R.R. which after it enters Inyo follows along the Owens River until they both come to Owens Lake, an alkaline body of water. It is about his river that I write to you.
>
> This river after it leaves the narrow mountain canon, runs through a broad and fertile valley for over a hundred miles. The first 20 miles of which is all or nearly so, in cultivation. . . . Indeed the people are very very proud of their little valley and what their hard labor has made it. The towns are all kept up by the surrounding farms. Alfalfa is the principle crop. . . . Cattle raising is a great industry.
>
> There has never been any capitolist or rich people come here until lately and all the farms of the Owens Valley show the hard labor and toil of people who came here without much more than their clothes. . . . Now my real reason for writeing this is to tell you

> that some rich men got the government or "Uncle Sam" to hire a man named J. B. Lippancott to repersent to the people that was going to put a large damm in what is known as Long Valley. . . . Imagine the shock the people felt when they learned when Uncle Sam was paying Mr. Lippancott he was a traitor to the people and was working for a millionaire company. The real reason for so much work was because a man named Eaton and a few more equally low, sneaking rich men wanted to get controlling interest of the water by buying out a few or all of those who owned *much* water and simply "Freeze Out" those who hadn't much and tell them to "Git."
>
> Now as President of the U.S. do you think that it is right? . . . Is there no way to keep the capitolist from forcing people to give up their water right and letting the now beautiful alfalfa fields dry up and return to a barren desert waist? . . . As you have proven to be the president for the people and not the rich I, an old resident who was raised here, appeal to you for help and *Advice* . . .
>
> So Help The People of Owens Valley!
>
> I apeal to you in the name of the Flag, the Glorious Stars and Stripes,
>
> Yours Unto Eternity,
> Lesta V. Parker
>
> (Qtd. in Walton 147)

Parker emphasizes the valley residents' pioneer, yeoman traditions, their work ethic, and the progress they had made carving an Eden out of the desert. Cattle raising occupied a central, honored place in the valley's lifestyle, and the alfalfa that supported it was grown on irrigated land. Lionizing these traits despite their water-intensive character buttressed the myth of the garden. Parker's plea, as well as those of her neighbors, failed in this case to sway the state or federal government primarily because it could not subsume the perceived greater good of urban expansion. Nevertheless, allegiance to the traditional western values espoused by Parker helped fuel the engine of Reclamation that drove the West for most of the twentieth century.

This chapter examines the early collision between rural and urban values and its implications for the Reclamation era. The conflict over the

Owens River embodies many of the underlying tensions of the time. Both sides (valley residents and officials of Los Angeles) wished to divert the river's water without regard for the potential ecological consequences. Each side also claimed allegiance to the American Dream. Despite promises of unlimited water and lip service to yeoman ideals, there was, in fact, limited water and strong sympathy for urban expansion. Preaching political progressivism, cities began asserting claims to the hydraulic wealth made available through reclamation. Rural agricultural regions vehemently opposed such claims.

The Ford, as a representative literary text of the period, demonstrates how the two sides attempted to exploit the myths that had become woven into policy. Owens Valley residents portrayed themselves as quintessential yeoman pioneers who were settling the West and making the desert bloom. Los Angeles, meanwhile, viewed itself as the next phase in the West's evolution. Los Angeleños claimed that the primitive pioneers must make way for the modern city, with its growing population and burgeoning water needs. The depiction of this conflict in *The Ford* highlights not just the regional collision of rural and urban values, but also Austin's own struggle to balance fealty to "western" ideals of reclamation and irrigation with her own environmental ethics. Much of Austin's writing (including several passages in *The Ford*) suggest what later came to be known as ecofeminism. Some of the novel's structural difficulties stem from Austin's inability to align her proto-ecofeminism with the Owens Valley's wholehearted embrace of Reclamation.

The Ford, both by itself and in the context of Austin's entire canon, provides an excellent opportunity to view the ongoing tension underlying Reclamation. As Austin declared in *Cactus Thorn,* "It's really a question of water. When you have water, you can have anything you want. Anything" (5). Both sides in the Owens Valley struggle looked to utilize all available water to fuel their production-oriented local economies. Their dispute lay less in ideology than in methodology. None of the participants questioned the efficacy of a spiraling dependence on a resource already in short supply, nor did they question the ideology underlying the term *resource.*[1] That debate had to wait for the middle of the century, when ecology came of age as a science and "environmentalism" in the spirit of John Muir gained newfound political strength.

Neither the novel nor the fundamental questions it raises can be adequately studied without some familiarity with the history of the valley and its hydrology. This chapter therefore begins with an overview of the region's larger social and political history and the rise of its conflict with Los Angeles. It then turns to *The Ford*, viewing it as an avatar of the regional conflict and of the unresolved political, philosophical, and ethical issues underlying that conflict.

A Valley Gone Dry

The passage of the Newlands (Reclamation) Act in 1902 instigated an era of unprecedented hydraulic expansion in the West. The prospect of federal subsidy for irrigation projects opened up enormous possibilities for agriculture, while the likelihood of increased availability of water also brought about exponential urban growth. Shortly after the act's passage, residents of the Owens Valley began angling for the creation of a federal reclamation district that would enable them to dam the Owens River and use the stored water for agriculture. As things stood, many of the valley's crops depended on the caprice of the weather. Meanwhile, the Owens River, fed by the runoff in the High Sierra, ran unused into the sea. Periods of sparse rainfall—typical for desert ecosystems—were common, while farmers and ranchers daily gambled their livelihoods on the chance of rain.

Reclamation districts seemed the perfect poultice for this chronic threat to farmers' livelihoods. With the prospect of government subsidies for large-scale water projects, horizons for agriculture and human settlement expanded commensurately. The Owens Valley seemed an ideal candidate for federal intervention since it possessed a large, underutilized river and a growing agricultural community. Advocates of the proposed reclamation district stressed the valley's frontier conditions and their own pioneer values, thus demonstrating hardship and the need for federal aid coupled with a commitment to hard work and the American Dream. In addition, their already substantial agricultural achievements (the valley was a productive, food-growing region) indicated a solid potential return on any investment in the valley's hydraulic infrastructure. That prospect underscored the region's compatibility with the Reclamation Act's aim of efficient growth.

Owens Valley residents were naturally very excited when a man named J. B. Lippincott arrived in 1904, seemingly under the auspices of the federal government, and began surveying the valley's hydrology. Lippincott had been an engineer on John Wesley Powell's Irrigation Survey and currently worked as a survey engineer for the Bureau of Reclamation. Unbeknownst to those in the valley, however, Lippincott also worked for the Los Angeles Department of Water and Power. Coincident with Lippincott's arrival, another man, Fred Eaton, appeared and began purchasing options on local ranches at prices considerably higher than market value. Eaton, a former mayor of Los Angeles, claimed he wanted to go into cattle ranching but could not decide where. He often let options expire while purchasing others at equally high prices. Valley dwellers were delighted to be making what seemed like easy money while awaiting the creation of their reclamation district.

Their contentment was short-lived. Lippincott turned out to be acting not in his capacity as reclamation engineer, but rather as agent for Los Angeles. Eaton, meanwhile, knowing of the city's plans for the river, had been purchasing options on those ranches with the best water rights. He planned to buy the ranches cheap from valley residents and then sell the water rights to the city at inflated prices. With all this transpiring in the valley, city officials (including Lippincott) had also been quietly lobbying the federal government to table the notion of a reclamation district and support the city's plan for the water instead.

In 1905, the *Los Angeles Times* broke the story beneath a headline replete with historical irony: "TITANIC PROJECT TO GIVE CITY A RIVER." The city government soon released details of its plan to build an aqueduct that would divert the Owens River to the city's reservoirs. Lippincott's services on the city's behalf earned him a "commission" of twenty-five hundred dollars—more than half his annual salary.[2]

Los Angeles officials had argued that the water could be put to better use by the city than by residents of the Owens Valley, and the federal government eventually agreed. With the completion of the Los Angeles–Owens Valley aqueduct in 1913, the Owens River vanished, and with it went the valley's agricultural base. Formerly irrigated pasturelands reverted to desert. It was just this specter of redesertification that valley dwellers had used to lobby against the aqueduct (Walton 145). To them, allowing a "reclaimed" agricultural region to return to desert ran counter to all logic and was tantamount to lunacy. Cedric Belfrage, in a novel of the Owens Valley entitled *The Promised Land,* highlights the insanity of such a policy: "I'd like to know . . . what kind of country this is. You work all your goddamned life to make a home in a goddamned wilderness, and then a few bastards . . . can come along and turn it back into a wilderness again and call it progress" (58). In the eyes of many westerners, diverting water from agriculture to urban expansion amounted to a capitulation to nature—an admission of insufficient water to fuel infinite growth. Federal and city authorities skirted this inconsistency in the myth structure by diverting what water there was to the supposed infinite expansion of Los Angeles. Allocating federal hydrodollars to one region rather than another thus became a matter of preference rather than exigency. Indeed, optimism regarding the city's continued development was what fueled the Owens River's appropriation in the first place.

At the time of the aqueduct's construction, Los Angeles had more than enough water to meet its needs. However, its population more than tripled in the 1880s and tripled again between 1900 and 1910. The city's campaign for the aqueduct arose from "a need founded in prospect" (Hundley, *Thirst* 152).[3] Planning for L.A.'s ascendancy permitted officials to view constraints on further water projects as economic (i.e., there was not enough federal money to finance both the valley's reclamation and the city's growth) rather than hydraulic.

Owens residents, outraged at what they perceived as the city's subterfuge, campaigned vigorously to revive the now moribund reclamation district. Their protests worked to little avail. President Roosevelt declared that "this [water] is more valuable to the people as a whole if used by the city than if used by the people of the Owens Valley" (qtd. in Walton 150). This concept of the greatest good for the greatest number underlay both the Reclamation myth and the production-oriented economy.

Garrett Hardin's seminal essay "The Tragedy of the Commons" refutes the logic of the greatest good for the greatest number, arguing that simultaneously maximizing for the greatest good and the greatest number is mathematically impossible. He uses the example of an unfenced cow pasture to show that as long as the net gain to a farmer of adding a cow is larger than the net loss of others adding cows—net gain is, in this case plus one cow, while the net loss is a fraction of the commons area—farmers will always seek to maximize their gains until the commons is destroyed. In using this example, Hardin undermines a proposal of Powell's, which advocates unfenced common pasturelands as a key to judicious land management (see chapter 1).[4]

Despite its logical problems, the principle of making the best use of available resources formed the plinth of the American conservationist movement as designed by Pinchot. Pinchot's maxim, "There are two things in the world: human beings and natural resources," plainly states his allegiance to the concept of human control and domination of nature. Conservation, as designed by Pinchot and espoused by Roosevelt, aimed to "control nature and serve the material interest of humankind but with an eye to long term needs" (Nash, *Rights* 9).

The Roosevelt administration believed that diverting the Owens River to Los Angeles would serve the long-term needs of a greater number of people. John Walton argues that the change in allegiance at the federal level from pioneer communities to larger urban developments reveals an important shift in policy. That shift reflected the new realities of a nation with a closed frontier. Pioneer society had initially benefited from the federal emphasis on conquering and developing the West. During the progressive era, however, federal priorities shifted to urbanization. In the Owens Valley, "this second phase . . . was as traumatic as the first, but its victims were now its former agents" (Walton 194).

The "theft" or appropriation of the Owens River to slake Los Angeles's growing thirst signaled a clear victory for the forces of progressivism.[5] Farmers and ranchers in the Owens Valley clamored that their livelihoods were being sacrificed at the altar of urban sprawl. City officials and media, meanwhile, depicted valley dwellers as ignorant bumpkins standing in the way of progress.[6] Valley residents felt betrayed; they had believed themselves the beneficiaries of a soon-to-be declared federal reclamation district that was to have turned the valley into an agricultural paradise. Instead, city agents duped them through a series of questionable dealings.

The aqueduct went on line on November 5, 1913, despite the concerted efforts of the local populace (including acts of sabotage—arguably the first organized acts of environmental sabotage or "ecotage" in the nation's history).[7] The Owens River disappeared from the valley, flowing instead into the faucets and lawns of Los Angeleños.[8] In the Owens Valley, sagebrush displaced alfalfa. Dust storms became common because the topsoil, first anchored by desert flora and then uprooted and planted over with crops, no longer had any anchor against the driving wind. These same conditions—land reverting to desert amid dust storms and disappearing topsoil—recurred a generation later slightly further east, causing the Dust Bowl.

Whereas the Dust Bowl that tormented the Plains states resulted from decades of land abuse and the looming dominance of factory farming (see chap. 3), the Owens Valley had not been long enough under the yoke of industrial agriculture for that to have caused its ecosystemic woes. Instead, its hydrology had been sacrificed to the ecology of the city. Residents of Los Angeles, meanwhile, dwelling in a city in the desert, not only felt the biological need for water, but also sought its power and wealth-signifying allure.

It bears reiterating that neither the city nor the valley was motivated by "environmental" concerns in the sense in which that term is used today. Each side saw the valley's water as a resource and source of potential wealth. Residents of the Owens Valley had no objection to the construction of a large diversionary structure that would forever alter the course of the river. They simply wanted a share in the economic windfall such a structure would create.[9] The restorationist grail of a precapitalist Eden would not become a factor in the land-use debate until the latter half of the century. For the moment, the conflict lay in the imminent unseating of agriculture as the federal favorite son.

Despite a continued allegiance to the yeoman ideal (as evidenced by the passage of the Newlands Act), a community could not ensure its status as federal beneficiary simply by virtue of a functioning agricultural apparatus. The cultural pull of the yeoman ideal and the myth of the garden remained formidable, but reclamation decisions were ultimately matters of monetary appropriation. And monetary decisions are rarely determined by sentiment. As Raymond Williams notes, decisions concerning the fate of competing urban and rural interests are the province of a powerful minority who control capital and determine its use according to calculations of profit (295).

In the Owens Valley, as in the West in general, the locus of the decision-making process lay in the use and control of water. Its dominion dictated not just wealth, but also the prevailing geography of the region. Dams and diversionary projects changed the landscape. Where there was water, there could be food, and people, and all the concomitant changes they brought. In the valley, water also carried the power to decide who would work the land and how.

While the aqueduct's local importance involved control over the Owens River, its national significance lay in the disputants' positioning within the American myth structure. The national resonance of their respective versions of the American Dream determined access to federal largess. The spoils were, of course, hydraulic.[10]

At both the local and national levels, the struggle centered not on what constituted a resource, but rather on who controlled it and how best to use it. Any unexploited resource was "wasted," subject to claim by anyone with the gumption to put it to use. This was especially true of water. To have access to water and not use it for personal gain denoted irresponsibility, making the water right subject to forfeit. This ideological underpinning for what amounted to a contest over a capital asset underscores what Jean Baudrillard, among others, cites as an egregious flaw in economic systems driven by the forces of production. To unthinkingly link humans with their capacity for production without first interrogating the motives for producing or the need to produce omits a crucial step in the socioeconomic analysis. That omission undermines both Marxism and capitalism alike.

Baudrillard believes that the Marxist ideal of "transforming nature according to human ends" shares a fundamental misassumption with capitalism. Both systems attach a perceived good to labor power. Marxism posits that workers are alienated by the sale of their labor power, whereas capitalism celebrates the benefits gained through its sale. Neither scheme allows for the more radical hypothesis that workers might be alienated *as* labor power and that transforming nature may not be a worthwhile goal (Baudrillard, *Mirror* 31). It matters little whether workers become alienated from the products of their labor and demeaned by forced participation in a system based on exchange value (as with capitalism) or are "liberated" from the wage-labor cycle yet subjugated by a system that measures human worth in productive capacity and output (as in Marxism). Under either regime, workers remain indentured to ever greater output and technological domination of nature.

This ever-worsening inequality among humans and between humans is evident in an increasing dependence on technology. Will Wright maintains that this prevailing obsession with technological domination and extractive methodology underscores a fundamental incoherence in a science-based cosmology. A society that characterizes technological advance as anything contributing to economic or personal short-term gain ignores the larger issues of sustainability and availability of supply (35). In theory, nature can always yield more. Relying on nature's unlimited productive capacity relegates concerns over ecology or supply to the realm of the technological.

It is precisely this allegiance to the exploitation of nature and to the human role as agent of production that Baudrillard cites as a central problem within the human condition. The Owens Valley conflict exemplifies the dueling agendas that characterize this mind-set. Los Angeles sought to fuel an ecologically unsustainable rate of urban development whereas the people of the Owens Valley looked to build an agricultural apparatus that would facilitate their participation in an ecologically devastating factory farm economy.[11] Both sides ignored long-term consequences, preferring to focus on the illusory surplus created by increased extraction of finite water resources.

Throughout the West, tremendous amounts of engineering know-how and capital went into maintaining the fiction that the water supply would never diminish as long as technology continued to improve the means of diversion and delivery. Fictions such as these arise from science's ability to place itself above those who would question its motives. By claiming objectivity, science appears to strip itself of ideology. Scientific actions that bolster the extant social structure attain a cloak of impartiality that immunizes them from social critique. This guise of objectivity perpetuates ecological devastation in the name of technological advance.[12]

"Good science," according to Wright, aspires to long-term ecological stability rather than short-term extractive gain: "The idea of rationality is about sustainability. . . . [I]ts inherent social criteria are the criteria of ecological coherence and reflexive evaluation, the criteria of legitimating sustainable institutions" (154). Just because technology enables the exploitation of a given resource does not necessitate or legitimate such a move. Technology is an agent of the status quo, producing tools that serve the needs of those in power. In a profit-driven economy, a project that does not enrich its sponsors is a bad investment, regardless of its ecological beneficence. As Andre Gorz notes, "Windmills . . . lost out because wind is everywhere, they couldn't be monopolized" (61).

In the West, technology permitted the proliferation of water-intensive developments despite their incompatibility with the region's hydroregime. Its scarcity combined with its biologically necessary role combined to make water the linchpin of the region's economy. Its exchange value became its defining characteristic, while the myth of water's abundance and the technology surrounding its procurement magnified the power struggle over its control.

An economic system dedicated to ever-increasing extraction makes humans and nature adversaries despite their being biologically intertwined. Such a system is illogical. According to Baudrillard, "What man gives of his body in labor is never given or lost or rendered by nature in a reciprocal way. Labor only arms to 'make' nature 'yield'" (*Mirror* 44).

This notion of a yielding (albeit indentured) landscape fits neatly with the feminization of nature that typically accompanies extractive economies and that is integral to the western myth system. Without ample water, reclamation would stall and the desert would remain "barren" and "infertile." This pejorative linking of female imagery (and, consequently, women) to land and to a subservient role in the social system forms part of a strategy of domination. Carolyn Merchant believes it forced both women and nature to become "the locus of the moral law and emotive expression," enabling a "mechanical and moral consciousness" that legitimated exploitative modes of production, reproduction, and ecology (*Revolutions* 232). Extractive technology abounds with sexual imagery, and this too is consonant with the land's dual image of mother and mate. In *The Great Gatsby*, for example, Jay Gatsby's pursuit of

Daisy maps on to the quintessentially American pursuit of "the fresh green breast of the new world." Just as Gatsby sought from Daisy to "suck on the pap of life, gulp down the incomparable milk of wonder," so too had generations of Americans attempted to possess the continent whose "vanished trees . . . had once pandered . . . to the last and greatest of all human dreams" (Fitzgerald 182).[13]

Linking women with the land and classifying both as reproductive agents and tools of the patriarchy allowed their use and abuse in the name of the common good. They become resources, and natural resources are, by definition, national assets. Managers exploit resources for the benefit of the nation. To do otherwise is wasteful. Often, the well-being of an object and the perceived national good come into conflict. Classifying the object (human or otherwise) as a resource negates this potential friction by equating its good with the well-being of those who control it. Just as women were expected to subordinate their needs and desires to the patriarchal ideal, so too did resources lose their individual identity upon becoming linked with the national good.

In the Owens Valley, the problem lay not in a failure to yield, but rather (metaphorically speaking) in a big guy muscling in on a smaller guy's girl. The region's virtue lay helpless before Washington's duplicity and L.A.'s groping embrace. A popular refrain in California opined that "the federal government held the Owens Valley while Los Angeles raped it." This phrase belies a highly important change in attitude, one that shifted the moral grammar of the conflict. Pioneers and yeomen deserved respect, but violated innocents *demanded* justice.

While the feminization of the land and the subjugation of both women and nature had been going on for much of recorded history, the struggle over the Owens River occurred at an important juncture for American geography. The Reclamation era had initiated an exponential increase in settlement west of the hundredth meridian. Shortfalls, corruption, and urban-rural conflicts were the inevitable result. Characteristically, the land, feminized by the language used to describe it, bore the blame for its supposed failure to yield. The land was deemed barren, in need of technological intervention to "restore" it to health. Without human intervention, the fickle land would revert to its previous "unproductive" state.[14]

As westerners strove to portray their respective regions as most congruent with the national vision of progress and the American tradition, Owens Valley residents found themselves hamstrung by a national swing toward urban progressivism. Though reclamation projects were legally required to benefit agriculture and irrigation, the federal government would not allow them to interfere with urban prosperity. In later years, reclamation projects came to dominate the western landscape by providing subsidized water and power to both urban and agricultural centers. These large-scale endeavors came to pass in part because state and federal governments learned from the example of the Owens Valley.

Austin's Ambivalence

Because of its pivotal place in American hydrological development and in California's history, the Owens Valley's tribulations have spawned a number of books, essays, and films (the most popular being the movie *Chinatown*). I focus here on Mary Austin's *The Ford* for several reasons. The novel sprang from the pen of one of the foremost nature-writers of the time as well as one of the country's first "environmentalists" and "feminists" (though neither term attained its current definition during Austin's lifetime).[15] Austin also lived in the Owens Valley during the period when Los Angeles began its quest for the region's water. In fact, her then-husband, Wallace Austin,[16] donated a water right worth over five thousand dollars to the ad hoc committee in the valley that had organized to fight the aqueduct. His action appears in barely disguised form in *The Ford* (387).

Austin had made her reputation as a "desert rat" and writer of western landscapes. Her earlier and most successful work, *Land of Little Rain* (1903), was a collection of vignettes about the California desert and those who dwell within it. Its evocative descriptions of the landscape and its inhabitants offered an early paean to the notions of environmental protection and limits to growth. Even the book's title suggests nature-imposed parameters on human activity in an arid climate. According to Lois Rudnick, Austin was one of the first to advocate a nonexploitative, harmonious relationship with the land:

> Austin . . . opposed the concept of the land as virgin and the Judeo-Christian imprimatur that man must rule, tame and reform "her." . . . That no one had mastered it—indeed that it could not be mastered—was attractive to [someone] seeking to make [an] impact on American literature. . . . Even when envision[ing] the land in specifically female terms, "she" is not typically exploited or exploitable. (11–12)

By the time of *The Ford* (1917), Austin's outlook had changed. Part of the text's forced happy ending involves the victory of Kenneth Brent, protagonist and displaced agrarian, over Old Man Rickart, venture capitalist. The valley's salvation lies, Austin suggests, in the ascendancy of Brent's view that those who work the land have the right to use it however they see fit and are entitled to whatever yield they can coax from it (*Ford* 436). To Austin, the valley's expanding agricultural aspirations were both natural and salubrious to the region's ecology, even if they did not jibe with her overall vision of land use.

The novel opens at the Arroyo Verde [Owens] River with a symbolic baptismal scene involving Kenneth Brent. The landscape is pastoral and, while dangerous (young Kenneth nearly drowns), strongly evokes the bucolic splendor that drew the Brents to the region. Nevertheless, Kenneth's father, Stephen, knows well that the agricultural community of Tierra Longa cannot survive without a dam to divert the river to their fields. To him, a dam is not only necessary, but natural. "Wasting" precious water flouts common sense and nature's will. According to the narrator in *The Ford*, "It was a little river, but swift and full, beginning with the best intentions of turning mills or whirring dynamos, with the happiest possibilities of watering fields and nursing orchards, but discouraged at last by the neglect of man, becoming like all wasted things, a mere pest of mud and malaria" (34). The narrator's claim that rivers (and all other things as well) require human intervention to avoid a sordid and ugly uselessness goes unchallenged despite its apparent incongruity with Austin's oft-stated reverence for nature.

While Austin was able to admonish in *The Land of Little Rain*, "The land will not be lived except in its own fashion" (33), and later (in *Stories from the Country of Lost Borders*) to note, "It takes man to leave unsightly scars upon the face of the earth" (88), she nevertheless felt that certain human uses could and should shape the land. In *Lost Borders* Austin contends,

> It is the proper destiny of every considerable stream in the west to become an irrigating ditch. It would seem that the streams are willing. They go as far as they can, or dare, toward the tillable lands in their own boulder-fenced gullies—but how much farther in the man-made waterways. It is difficult to come into intimate relations with appropriated waters; like very busy people they have no time to reveal themselves. One needs to have known an irrigating ditch when it was a brook and to have lived by it. (123)

Here, Austin displays a marked ambivalence toward the channelization of rivers even while acknowledging it as their destiny. One cannot become familiar with a channelized stream—one needs to have lived by it in its previous incarnation. Yet if the streams' telos is fulfilled, no one will know them. For a desert rat like Austin, such a fate is usually unthinkable. Again, Austin's reluctant progressive bent collides with her nascent ecological worldview.

The result, at least in *The Ford,* is a clear victory for progressivism. Stephen Brent, depicted in the novel as a gentle man of the earth, is also the father of the valley's reclamation project. He adds rhetorical force to his vision by linking it to history and the destiny of the region's inhabitants:

> "Fifty years ago, when my father drove over the Pass, there was this house here and hacienda at Agua Caliente. . . . Even in those days they dreamed of a dam at Indian Gate. . . . Water," he said, "water and power . . . and farms . . . farms, not cities." (361)

Brent's willingness to exploit the land to suit human desires is evident throughout the novel and nowhere more obvious than in the following: "Wherever the land flings a handful of corn we run and scramble for it like beggars in the street. And she laughs—she laughs. I tell you, Burke, we've got to master her—we've got to compel her" (62). This overtly adversarial stance belies Brent's gentle agrarian exterior. "Mastering" and "compelling" the land form the basis of an extractive economy dedicated to development irrespective of ecological parameters—precisely the offenses for which Los Angeles stood accused.

Austin nevertheless characterizes the dispute as one between a business proposition and a way of life (367). She ignores the fact that the region's desire for a reclamation district stemmed from its residents' vision of the valley's becoming, through agriculture, a more productive participant in the nation's extractive economy. Even as Tierra Longa's inhabitants attempt to distinguish themselves from the faceless capitalist invasion by trumpeting their traditional values and land use, their rhetoric betrays them. They envision themselves as true yeomen living close to, and off of, the land, but they participate in the same extractive economy as the city they fight.[17]

Their behavior buttresses Baudrillard's contention that the "mirror of production" allows no other conception of nature than as the dominated other (*Mirror* 58). He argues that humanity is incapable of imagining nature as anything other than a vehicle for human needs. We are trapped by the constraints of economy and language—neither of which has the flexibility to shape another way of seeing the nonhuman environment. "Nature is the concept of a dominated essence and nothing else. . . . And it is by being sublimated and repressed that Nature becomes a metaphor of freedom and totality. . . . Everything that invokes Nature invokes the domination of Nature" (55–56). Similarly, even while celebrating the unconquerability of the desert, Austin advocates its subjugation and development. She appears to half agree with Powell that human activity can do no lasting damage to a planet that is so vast and apparently impervious to harm. Yet in other contexts, Austin decries the "scarring" that humans wreak upon the earth.

This dichotomy may be a product of its time. In 1878 in *Report on the Lands of the Arid Region,* Powell could not imagine any human activity that could cause terrestrial harm of any significance. However, by the time Austin published *The Ford* in 1917, human capacity to cause harm had grown exponentially. The era of total war had arrived, bringing with it trench warfare, mustard gas, the machine gun, and aerial bombardment. Technology's capacity to destroy both people and the land was starkly and permanently revealed.

Austin recognized the revolutionary implications of recent technological breakthroughs but retained a strong allegiance to an earlier time when the land's permanence provided comfort and certainty. The myth of the garden played off that ambivalence, offering a seductive fiction that humans could redesign nature and bring the desert to bloom while allowing adherents to believe that their actions did no lasting harm. Thus could Austin advocate the channelization of rivers, secure in the knowledge that the changes were for the good and that the land was too vast and the region too dry for humans to alter it in any significant way.

By contrast, Los Angeles's appropriation of the Owens River qualified, in Austin's view, as wanton environmental destruction and a human tragedy. Late in the novel, when the loss of the river seems a fait accompli, the narrator laments the seeming inevitability of the valley's defeat. Elwood, the fictionalized Fred Eaton character, and the conniving urban capitalism that he represents will ultimately defeat the people of the valley because Elwood represents "progress": "The difference between what they fancied lay behind Elwood's schemes and what . . . was Elwood's likeliest motive, gave to inevitable defeat the quality of ancient tragedy; the tragedy of men defeated, not squalidly by other men, but by forces within themselves which had the form and dignity of gods" (*Ford* 290). Austin implies that the valley residents cannot win because their opponents spring from within some inherently more powerful part of themselves. This is a telling passage; it acknowledges that the same mentality that created the concept of Reclamation created the propensity to urban sprawl as well. The two phenomena are intertwined. Demand for crops comes from those who do not grow them. Nonfarmers tend to live in urban areas, and it is they who provide the market for the crops grown in reclamation districts. The symbiosis would be beneficial for all parties but for the inevitable competition for the region's scarce water. The disparity in water wealth inevitably leads to an inequality in capital accumulation as well.

Earlier in the novel, Austin ironically notes the irresistible allure of "Capital" on the valley dwellers and their steadfast belief that if they too had it, their rural vision would hold sway. The irony, she suggests, lies in their ignorance of the true nature of capital. They cannot control it because capital, like water in the West, flows toward the rich and powerful. One acquires capital by participating in precisely that economic system currently driving the valley to ruin. One cannot be a rich yeoman because one becomes rich on the backs of yeomen.

Austin describes with little sympathy the futile quest for wealth of the men of the valley:

> They must have Capital; and if they did, such was their deep conviction Capital in the end would have them. Capital went about seeking whom it might devour, yet such was their strange illusion about it that they believed that if once they could lay hands on it, Capital could be made to run in their harness, breed in their pastures. To those who owned Capital . . . it ate out of their hand, but its proper nutriment was the contents of poor men's pockets. (*Ford* 95)

The people of the valley believe that they can make capital join nature as a slave to human ambition. Yet Austin harbingers Baudrillard by suggesting that no economic system driven by exchange value can function as anything but an engine of production that will ultimately impoverish the vast majority of its functionaries, including nature. Nonetheless, she makes an impassioned plea not just for the valley's rescue, but also for the establishment of a reclamation district. Austin's dilemma embodies the contradiction at the heart of early-twentieth-century environmentalism. Suspicious though she was of "conservation" as espoused by Pinchot and Roosevelt, Austin could not escape its grasp.

The Ford also draws an interesting parallel between the water war and the petroleum boom that took place roughly conterminously in California at around the turn of the century. As depicted in the novel, venture capitalists first invaded the valley for its oil, making a healthy profit for themselves at the expense of local inhabitants. Then, just as the community began to rally around the notion of a community-based irrigation system, the capitalists reinvaded, seeking to co-opt that resource, too. Water and oil are linked capital assets—commodities whose control carries the power to shape the local geography.[18]

In an economic system designed to allow those with capital to continue amassing more, control over important commodities naturally falls to those in power. Yet, such is the case with the valley's petroleum, with its water soon to follow. Yet, despite the dismal fate suffered by the actual valley, the novel, like most modeled on the Owens Valley, has a happy ending. This dogged optimism, persisting despite the actual loss of the

Owens River, illustrates both the bullish enthusiasm of the Reclamation era and the potency of its accompanying myth system.

In most ways, *The Ford* is a typical melodrama. The Brent family and their neighbors live a debt-ridden but happy existence in the Inyo (Owens) Valley. The only discordant note comes from Marcia Brent, the woman of the house. She longs for the culture and clean floors of city life and considers every day spent in Tierra Longa as another step toward martyrdom. Eventually, after pressuring her husband into an ill-fated oil venture and seeing it fail, Marcia Brent falls ill and dies. Husband and wife share a moment of redemptive bonding right before she goes to her reward. With his wife's death, Stephen Brent's zest for living dies as well. He becomes lethargic and all but unable to care for his two children.

The children mature rapidly. Soon, they are the caregivers. The family continues to scrape a meager living from ranching while gazing covetously at the Arroyo Verde (Owens River) and longing for a dam to divert its waters to their fields. They fantasize about the beautiful landscape that the dam will reclaim for the valley: "Once the gate was shut again by a concrete dam, Tierra Rondo would resume its ancient use as a lake bed for the stored waters of the river, which now ran all at large in a wide shallow forge far seaward, and in wet years mingled its freshness with the tide" (298). The dam is the particular dream of Stephen Brent. His fantasy of a dam in the valley coincides with what Austin suggests is nature's original intent. Building a dam becomes a natural act that will fulfill the thwarted designs of nature. This reasoning allows Austin to skirt the issue of human defilement of the land and instead construct a scenario wherein yeomen like Brent "restore" the land to its intended use. "Restore" thus becomes conflated with "reclaim."

The telos of Reclamation required large-scale dam-building and expenditure of capital and resources to support the fiction of a terrain and climate conducive to agriculture. Austin's attempt to link the concepts of restoration and reclamation indicates a need to rationalize the subjugation of nature. Even she, who loved the desert and its aridity (she wrote lovingly of Death Valley, among other places, in *The Land of Little Rain*), harbored the notion that the West could and should support expanded cattle ranching and agriculture. But, as the true-life Owens Valley example shows, that vision of western life had to contend with the looming specters of the city and venture capitalists. Urban capitalists sought to extract surplus value while ignoring entrenched cultural traditions. Rural people formed easy prey for false promises of wealth and devised schemes to enrich themselves without regard for yeoman traditions.

In Austin's Tierra Longa, the first temptation lies in oil. The petroleum venture fails because Stephen Brent and his partners attempt to lock horns with Old Man Rickart, symbol of corporate capital. Kenneth Brent would later understand the magnitude of his father's error: "He would see in a flash, Rickart and his kind as a work of nature, gigantic, inevitable, like the epoch which stored the oil under Petrolia" (*Ford* 245). Portraying a venture capitalist as a "work of nature" even as he seeks to destroy the valley demonstrates the depth of Austin's ambivalence regarding humanity's role in the biotic economy. Her ambivalence echoes that of Frederick Jackson Turner, whose seminal essay of 1892, "The Significance of the Frontier in American History," sought to situate western expansion within the larger fabric of American culture. Turner argued that the wilderness was necessary to the formation of the American consciousness. Pioneers settled the wilderness, and civilization followed on their heels. The newly civilized society then marginalized pioneer values, causing the pioneers to move west and the cycle to begin anew. In this vision of American culture, pioneers are simultaneously the cutting edge of culture and an anachronism. In many ways, Rickart is a metaphor for Turner's vision of the West's evolution. As a venture capitalist born in Tierra Longo but now allied with the city, Rickart embodies Turner's thesis that pioneers simultaneously represent the apotheosis of the human condition and an intermediate stage in the progression toward an urban existence.

Rickart seems to have an interest in every profit-oriented venture and no compunction about ruining anyone—even friends and neighbors—who interferes with him.

> Lands, water and minerals, he took them up and laid them down again, wholly uninformed of the severances and readjustments made necessary by that temporary possession. . . . [H]e did not know that men are warped by these things and that women died of

> them. It was as if a huge bite had been taken out of the round of his capacity, and left him forever and profitably unaware of the human remainder. (176)

Rickart's business savvy and his unconcern for others earn him the grudging admiration of those in the valley. They want to be like him even as he tramples them underfoot. Opposing him in a business venture is the commercial equivalent of defending the Alamo. The valley's population learns this again through Brent's dream of irrigating the area.

Young Kenneth Brent, hoping to learn how to survive in the business world, goes to work for Rickart, while his sister Anne, by far the most intelligent and interesting character in the novel, becomes a real estate broker in the valley. Anne plans to develop the region in the image of her father's dream even as Rickart mounts an underhanded campaign (using Elwood) to garner the valley's water and sell it to Los Angeles. At this point, the novel's plot roughly follows the real-life struggle between the city and the valley. In the novel, Kenneth Brent quits his position with Rickart and dedicates himself to saving the valley. He had inherited his father's passion for reclamation, and the shared commitment between father and son forges an almost transcendental bond between them. Austin suggests it forms a similar bond between them and the land: "the figure of the watery waste spelled much to his father, that in the moments when they faced it together . . . seemed wondrously shared between them" (34).

Kenneth Brent has an epiphany when he realizes that the Rickarts of the world do not share his feeling for the land. The father-son bond of the Brents contrasts with that of the senior and junior Rickarts. The younger Rickart and Kenneth have been friends since childhood, but their respective conceptions of Tierra Longa mirror the differences between yeoman traditions and corporate commerce: "[T]he long, lion-like valley which still held for Kenneth the only sense of home, of the continuity of existence, was to his friend one of the pieces of the game. The river with its fruitful possibilities he would have wound up like a scarf and tossed to one side or the other as the play went" (165). Even in this brief passage, the balance of power is clear. Brent's worldview and happiness pivot upon the caprice of Rickart and other venture capitalists. Rickart and his allies own much of the valley and have no compunctions about desiccating it for profit. This mercenary outlook gains them a great deal of leverage, leaving Brent with no reciprocal purchase.

The plot of *The Ford* then veers from true-life events. Anne Brent's plan to wrest the valley's water back from Rickart and the looming urban monolith succeeds, thanks to Kenneth Brent's labors. Los Angeles, in the guise of Rickart, decides to get its water elsewhere, ostensibly because Rickart "didn't really want the water of Arroyo Verde" and was "sore at having his plans interfered with" (422). Anne professes a love she does not feel for Rickart's already betrothed son in order to heal the rift between Kenneth and the elder Rickart. Kenneth marries the woman Anne had chosen for him while she (Anne) retreats into a sage and sanguine spinsterhood. The valley is saved; water will flow into local irrigation ditches rather than the city's faucets. The yeoman forces of reclamation have stymied the urban armies of progressivism, enabling the river to once again serve the needs of agriculture rather than those of the city.

Of course, the Owens Valley's actual history differs starkly from Austin's version. Over the more than eighty years since the completion of the aqueduct, the valley has evolved into a national symbol and rallying force against urban imperialism. Ironically, the loss of its water meant that the valley was never developed. Rather than turning into an agribusiness hub like the Imperial Valley, the Owens Valley gained local renown as an uncrowded wilderness area. But in the 1970s, Los Angeles began massive groundwater pumping in the valley, destroying the natural vegetation and drying up remaining springs and artesian wells. Valley dwellers and environmentalists joined forces for a series of court battles. There were also once again sporadic acts of ecotage and property damage, including the shooting of an arrow attached to a stick of dynamite into the William Mulholland Memorial Fountain in Los Angeles.

By the early 1980s, when it became clear that Los Angeles was losing in both the courts of law and public opinion, the city began negotiating in earnest for a mutually acceptable settlement. Negotiations continued throughout the 1980s, and in 1991 the parties reached a settlement. The agreement required Los Angeles to rewater fifty-three miles of the "dry channel" of the Owens River, establish off-stream ponds and wetlands, and undertake other environmental projects. The city also had to pay Inyo County $2 million annually to offset the county's reduced tax base stemming from Los Angeles's ownership of its land. The agreement also made the city's groundwater pumping subject to environmental considerations. While substantial, the Owens Valley's victory was not complete. Among other problems, residents remained unsatisfied that the city did not have to rectify its desiccation of the Owens Lake. The dust from the dry lake bed continues to generate choking dust storms that sometimes carry all the way to Bakersfield, a distance of roughly seventy miles.

Austin's fictional solution to the valley's water crisis concerns itself more with catharsis than with the situation's actual complexity. However, *The Ford* does anticipate ecofeminism by linking the subjugation of women with that of the land and merging the two into a potent force for social change. Anne's role in the valley's salvation offers an excellent view of Austin's early feminism as well as her proto-ecofeminism. For example, Kenneth Brent is essentially a dolt. Without Anne's maneuvering and prodding, he never could have kept Rickart from the valley's water. Anne, for her part, understands that the preeminence of the patriarchy relegates her to a supporting role despite her superior business sense and intelligence. Even Kenneth realizes (with Anne's help) the congruence between the current system of land use and women's role in society: "Anne had called it a man-made world, and all he had seen of it went to show that men had made it badly. But what could men do in a world which land, waters, the worth of women, had no measure but a man's personal reaction (373). This passage groups land, water, and women—three potent signifiers in the western value-system. It also underscores their subjectivity. The exchange values of all three are, according to Austin, determined by the dominant ideology—in this case, capitalist patriarchy. That act of valuation has become involuntary, a rote practice descended from the original arranging of the signs in such a way as to perpetuate the (de)valuation of women and nature.

Anne recognizes the impossibility of openly flouting the power structure and chooses instead to subvert from within. She refuses to recognize the legitimacy of an economy organized around an arbitrary grouping of signs.

> Anne was not troubled by any incubus of a System. "There isn't any such animal," she insisted. "It's the way we look at things. . . . Society is a sort of mirage, a false appearance due to refraction . . . because we have an androcentric culture." (233)

Her philosophy contains much that would later become ecofeminism. She labels extant reality "man-made" while noting alternative ways of seeing that do not privilege men and invite the ascension of women.[19] She further contends that the male way of seeing the world, based as it is on an elaborate, constructed vision designed to gratify male impulses, is less realistic than the female perspective, since women see from the point of view of the subjugated: "Take marriage, for instance; a woman will marry a man because he is clean and honest and will make a good father . . . but a man won't marry a woman unless she makes him feel a certain way . . . unless there's a—mirage" (233). In Austin's view, women look for qualities that will facilitate healthy relationships and propagate the species. Men insist that women become accomplices in their fantasies.

Here too, Austin presages many ecofeminist concepts. According to many ecofeminists, women, like the land, serve as pleasure vehicles in a male-constructed Xanadu. Because land has traditionally been feminized, just as women have been objectified, the link between women and nature is stronger than that between men and nature.[20] Indeed, men have consciously alienated themselves from nature, choosing instead to link themselves with culture. The resulting dualisms of women/nature, men/culture, while traditionally serving to subjugate women and nature, can potentially empower women and provide the means for constructing a nonandrocentric society.[21]

Unfortunately, Austin's early ecofeminist awareness collides with her strong allegiance to Reclamation ideology and its gospel of terrestrial and hydrological domination. Her entire oeuvre demonstrates a continuing struggle to accommodate Reclamation with environmentalism. *The Ford* provides perhaps the best example of their incompatibility.

Austin believed that Los Angeles's appropriation of the water affronted the region's terrestrial integrity even as she endorsed the valley's plan to dam the river and force the water into irrigation ditches and fields that had been shorn of their native flora. She somehow managed to accommodate this land use with a belief that the land should not be lived "except in its own fashion" and with her smoldering anger over human scarring of the landscape. In *The Ford,* it is Anne who engineers the valley's victory over the city. She, who had decried the male-dominated world that constructed "mirages" that measured the worth of women and land solely in terms of their ability to offer gratification, uses her femininity to trap the younger Rickart into backing off the city's project. Her goal is to permit her father and brother to "master" and "compel" the river.

The conflicts within Austin's protoenvironmentalism exemplify the inherent tension infusing the myth of the garden. The land was supposedly Edenic, capable of supporting millions in bucolic splendor yet simultaneously in need of technological intervention to subdue it and force it to yield. Reclamation offered a compromise between these two poles. By "reclaiming" the land from aridity and "unproductivity," American diligence would revive a previous era of Edenic abundance. The land's health was tied to a commitment to its cultivation. Reclamation thus legitimated large-scale agriculture through its linkage with yeoman principles of land use and the biotic integrity of the ecosystem.

One of the final passages of *The Ford* offers an excellent encapsulation of Austin's outlook and a clear view of its close connection with the myth of the garden. In explaining to Rickart why he had flouted Rickart's authority and saved the valley, Kenneth Brent "make[s] clear his conviction that the earth was the right and property of those who worked it, and that its values should accrue to them if to anybody" (436). Thus, even Austin, one of the nation's most ecologically minded citizens, could not separate herself from the commodification of nature inherent to productive economies. As Anne says in the novel,

> Look at the land . . . the first thing to learn is that you can absolutely find out what land is good for, and in time we'll find that, no matter what you feel about it, it only belongs to the people who can do those things. . . . I can make a Socialist out of a prune man . . . by keeping him six years on a piece of ground that was meant to grow potatoes. (234)

Austin recognized that all visions of the land were ineluctably linked with production, and, despite misgivings about the ramifications of such a tie, she embraced it for lack of a better option. Of the two alternatives—yeoman husbandry and urban sprawl—Austin allied herself enthusiastically with the former.

This faith in the right of yeomen to shape and subdue the land, while at the same time condemning those qualities when evinced by others, portends the conflicts in *The Grapes of Wrath.* The Okies in Steinbeck's story also claim allegiance to traditional yeoman values while at the same time participating in ecologically disastrous factory farming. As in *The Ford,* water forms the privileged sign; its control determines who will work the land and in what manner.

Colorado River, looking upstream toward Boulder Canyon, future site of the Hoover Dam, circa 1928. (Courtesy of the Henry J. Kaiser Collection, Bancroft Library, University of California, Berkeley.)

Hoover Dam, construction in progress, circa 1933. (Courtesy of the Henry J. Kaiser Collection, Bancroft Library, University of California, Berkeley.)

Sharecropper entering California, circa 1935–45. (Farm Security Administration photograph, courtesy of Bancroft Library, University of California, Berkeley.)

Hoover Dam, Arizona-Nevada border. (Photograph by Robert Dawson.)

Owens Valley water leaving Owens Valley and entering Los Angeles. (Photograph by Robert Dawson.)

Drying Rye Patch Reservoir, Nevada. (Photograph by Robert Dawson.)

Spillway, Lake Berrysea, California. (Photograph by Robert Dawson.)

Dry Arkansas River used as a wheatfield, Dodge City, Kansas. (Photograph by Robert Dawson.)

3

Turning Wine into Water: Water's Crucial Absence in *The Grapes of Wrath*

Eastward I go only by force; but westward I go free.
—Henry David Thoreau

The Old Testament describes wilderness as "a thirsty ground where there was no water." When the Lord wished to punish, he threatened to "turn the rivers into islands and dry up the pools and . . . command the clouds that they rain no rain upon it." When granting redemption in Isaiah, God promises instead that "waters shall break forth in the wilderness and streams in the desert" and that "the desert and dry land shall be glad" (Deut. 8:7, 15, Isa. 5:6, 35:1, 6, 43:20).[1] The Garden of Eden provided the antithesis of desert wilderness, a place where water flowed freely and manna of all sorts lay ready to spring out of the ground. This is the imagery that spawned the myth of the garden. It is also one of the many biblical cross-currents running through John Steinbeck's canon. At the dawn of the common era, John offers Jesus his baptism in the river Jordan. Two millennia later, Jim Casy baptizes Tom Joad in an irrigation ditch.

The Grapes of Wrath represents an indictment of the Myth of the Garden and its accompanying myth of the frontier. The lever with which Steinbeck pries apart and ultimately dismantles these fictions is a critique of the agricultural practices that created the Dust Bowl and then metamorphosed into a new set of norms that continued to victimize both the land and its inhabitants. Both nineteenth-century homesteading (based on the Homestead Act of 1862) and agribusiness, its twentieth-century descendant (born from the failure of the Homestead Act), relied on the (mis)use of water to accomplish their respective goals. And both policies resulted in ecological disaster.

The Plains were called upon to supply grain for the international war effort in 1914 and to feed a hungry nation whose population continued to multiply exponentially. Throughout the nation, industrialization held sway as the isolationism of the nineteenth century gave way to the globalism of the twentieth. These transitions required great expenditures of resources and, in the Grain Belt, the resource most in demand was water. As farmers poured their short-term profits back into land and seed, their fates became ever more dependent on the availability of water. This era of expansionism coincided with a period of greater than average rainfall in the region. When the climatic pendulum swung back toward aridity, Plains farmers had to declare hydrological bankruptcy. Still, neither the farmers nor the federal government would abandon the myth of the garden. As the government scrambled to dam rivers and force water into the arid lands, farmers clung fast to their vision of uncountable abundance amid a green world.[2]

The class stratification depicted in *The Grapes of Wrath* arose from corporate control over water. However, the region's aridity made water an absent signifier. Both in the novel and in the desert itself, water's conspicuous absence is what makes it so powerful. The flooding that climaxes the novel is thematically situated to provide maximum counterpoint to the drought that originally forced the Joads to migrate west. Disenfranchised and dehumanized, the Joads can only curse the rising floodwaters even as they once prayed for a deluge to feed their parched crops.

By the end of the novel, the cycle of alienation is complete; people whose humanity was once integrally tied to the land and the weather now care nothing for the growing season or the health of the earth. Their survival has come to depend on shelter from the elements rather than the elements themselves. They have become components of the factory-farming process, economically distant from the corporate growers who oppress them, but closely tied to the industrial ethos that rewards the subjugation of nature. The primary difference between the growers and the migrants now lies in their respective relationships with water, the privileged sign. The growers, owners of the irrigation channels, centrifugal pumps, and watertight mansions, control it, while the Okies, starving and drenched, are at its mercy.[3]

In *The Grapes of Wrath,* Steinbeck presents an archetypal Plains family caught in the modernization of the American dream. Adapting to the realities of a closed frontier and a desert in the country's midsection involved retrofitting the dominant myths to encompass corporate capitalism as well as accepting water's scarcity and preeminence as commodity in the West. This shift in ideology completed the antiquation of the Joads' way of life. Ecological realities had long ago proven their lifestyle quixotic, but it took the coincidence of the Dust Bowl and corporate agribusiness to dislodge the Okies from their land and homes.

Steinbeck acknowledges water's primacy in the West by documenting the social ramifications of the ideology that permitted its monopolization and waste. The ecological decimation of the Plains, the preeminence of agribusiness, and the migrants' brutal poverty in California are all attributable to this hybrid of yeomanism and hydrocapitalism. Steinbeck's abiding affection for the yeoman agricultural ideal also appears throughout the novel. Donald Worster believes that this nostalgia comes at the expense of a coherent critique of the water-based oligarchy primarily responsible for the ecological degradation of the Southwest and its accompanying human suffering (*Rivers* 229). While Worster's criticism has substantial merit, it fails to address the symbolic power attached to water that pervades the novel. That in *The Grapes of Wrath* Steinbeck chose to stress his affection for the yeoman tradition rather than explicitly condemn modern hydraulic society does not detract from the book's success in subverting the hydroindustrial paradigm. From the drought in Oklahoma, to Noah's refusal to leave the river in Arizona, to the raging floodwaters that climax the text, Steinbeck weaves water into the novel's structure and also into virtually every thematically significant event in the novel.[4]

The reactions of the state and federal governments to the book's publication as well as that of the oligarchy-controlled California media demonstrate the novel's effectiveness. The migrants' struggle became a national cause célèbre, and the novel's verisimilitude was debated at the highest levels of government.[5] Vehement condemnations of the book and its author followed shortly after publication in 1939 and continued for years. The most vociferous denunciations came from the water barons and their political allies, suggesting that, contrary to Worster's contention, Steinbeck did indeed understand the politics of water use and that his novel attacked it successfully.

One of the most effective techniques used by the press to discredit the novel involved letters to the editor from supposed "Okies" protesting that the conditions depicted in the novel did not really exist. The letters told of friendly treatment by the growers, clean living conditions, and enough work for everybody. The papers also spread rumors of Okies wanting to kill Steinbeck for telling lies about them. Little information defending Steinbeck's version of events reached the public at large until a number of other exposés (most notably Carey McWilliams's *Factories in the Field*) were released and photographs documenting the migrants' conditions gained widespread notoriety.[6]

The first part of this chapter examines the complicated history of the region, offering *The Grapes of Wrath* as a lens through which to view the ecological, hydrological, and social factors that caused the Dust Bowl and the subsequent mass migration of the "exodusters" to California. The second section examines water's role as absent signifier more closely within the novel and relates it to contemporary difficulties stemming from water's use and control in the Southwest.

Creating the Dust Bowl

Not just in the Hebrew Bible but throughout history, the habitability of any region has been determined by the availability and accessibility of its water. The Spanish explorers who first traversed the Southwest deemed it an inhospitable wasteland, unfit for human settlement except by savages content to scrape their existences from the unforgiving rock. American trailblazers including Lewis and Clark and Zebulon Pike also held little hope that the arid region could sustain American settlements. Such criticism, however, quickly disappeared in the storm of patriotism that surged through the new United States. Parallel visions of world dominance and transcendental bonding with nature created a unique blend of ideologies that sought to sustain simultaneously an extractive economy and an unspoiled, untrammeled frontier. Not till near the turn of the twentieth century did the inexorable collision of these visions loom close enough to draw the notice of the nation's policymakers. This collision resulted in what Carolyn Merchant calls an "ecological revolution." Ecological revolutions are caused by increasing tension between ecosystemic requirements and the modes of production that drive the economy. When this tension reaches unsustainable levels, it generates "transformations in consciousness and legitimating worldviews" (*Revolutions* 5). The Newlands Act, with its implicit recognition of the hydrological parameters of the West, represented one such transformation.

In the Owens Valley, the tension that spurred the transformation took the form of a clash between urban progressivism and rural populism, with both sides asserting entitlement to the region's water. Urban forces prevailed in the valley, but elsewhere, in the agribusiness-controlled regions of California, agriculture remained dominant. The growers' control over water gave them virtually unlimited power. Later, when the exodusters flooded into the region, they quickly learned the scope and implications of the growers' primacy. The hydrological basis for the growers' power probably did not surprise the migrants, who were themselves intimately familiar with water's power to shape lives. Their very presence in California stemmed directly from a cyclical shortfall of water in the nation's midsection.

American history shows that people traditionally migrated to the Plains during periods of high rainfall. When the rains subsided to typical levels, people retreated or pressed on. But, by the 1920s, the frontier was closed and Americans had bought into the notion that technology and God would see to it that the Great Plains became the world's agricultural capital. They clung to this notion despite the fact that meeting the grain demands of a global economy in a region where annual rainfall fluctuated between seven and twenty inches made little ecological sense. When they could not meet their goals, Plains dwellers lashed out at the weather, believing it caused their woes. There was not enough water, they complained; the weather had failed them.

This argument is analogous to blaming a gravel pit for yielding no gold. That is not to belittle the very real human tragedy of the Dust Bowl nor to deny the nobility of many of those who suffered through it. Nevertheless, the Dust Bowl's ecosystemic catastrophe was both avoidable and remediable, though not by means that were palatable to the region's residents. They remained unwilling or unable to face the ecological realities of the region. According to Worster, typical Plains farmers of the time can be expected to "fail to anticipate drought, underestimate its duration when it comes, expect rain momentarily, deny that they are as hard hit as outsiders believe . . . admit that some help would be useful, demand that the government act and act quickly . . . without strings . . . pooh-pooh the need for major reform . . . eagerly await the return of 'normalcy'" (*Dust Bowl* 28). When New Dealers attempted to innovate, they met with entrenched hostility. Plains farmers believed that Providence would see to the fate of the region and "Providence, not Washington, would see them come out all right" (28). It is precisely this stubborn adhering to traditional values while implementing ecologically pernicious agricultural methods that brought on the "dirty thirties."

The values themselves have their genesis in the dual myths of the garden and the frontier. The Joads' saga in *The Grapes of Wrath* offers a fictional version of the consequences of those myths. Both derive from a perceived superabundance of resources, a national fantasy that prodded the Joads toward Oklahoma and then later to California. Belief in an infinite national trust fueled dreams of individual wealth and national dominance amid a rugged land that would never cease testing all those attempting to wrest an existence from it.[7] West of the ninety-ninth meridian, water's perennial scarcity threatened to undermine this popular vision. Rather than subvert the prevailing value system, Americans bought the notion of hydroabundance promulgated by ideologues like William Gilpin. Their insistence that rain followed the plow and their boasts that the West contained infinite supplies of minerals and timber convinced people like Grampa and Gramma Joad to move west, settle in the arid region, and take up the Jeffersonian yeoman ideal.

Jefferson, however, lived in Virginia. His philosophy stemmed from his intimacy with farming practices in the humid region. He was profoundly ignorant of agricultural techniques west of the Mississippi, and his ideas did not adapt well to the arid West. A century after Jefferson's time, Powell labored for close to thirty years to bring western land-use policies closer to western terrestrial and hydrological realities. By the turn of the twentieth century, the regional and federal governments acknowledged that agricultural practices in the arid lands required severe retooling. By then, however, powerful corporate interests dominated the region's economy. The conflict between misguided government policies, yeoman land-use ideals, and geographical realities had been exacerbated by the profit-centered tactics of agribusiness concerns.

Early in the novel, Steinbeck establishes the fundamental conflict between yeoman farmers and the banks over title to the land. According to a nameless tenant farmer speaking early in the novel,

> Grampa took up the land, and he had to kill the Indians and drive them away. And Pa was born here, and he killed weeds and snakes. Then a bad year came and he had to borrow a little money. An' we was born here . . . our children born here. And Pa had to borrow money. The bank owned the land then. . . . Sure cried the tenant men, but it's our land. We measured it and broke it up. We were even born on it, and we got killed on it, died on it. Even if it's no good, it's still ours. . . . That makes ownership, not a paper with numbers on it. (*Grapes* 34–35)

This passage reveals several of the guiding principles governing life in the Plains. First, the term "bad year" refers to inadequate rainfall and an accompanying water shortage, a cyclical reality of Plains life that formed one of the bases for the collapse of the yeoman life. Second, right of ownership was established through displacing Native peoples. That act, in and of itself, constituted (in the farmer's eyes) a right of title. Last, birthing and dying on the land created a blood-right of succession that no financial transaction could negate. The quotation reveals the teller's sadness that the laws of the country conflict with the laws of the land.

For Plains farmers, working the land was the litmus test of possession. Agrarian ideology held that only those who work and love the land can truly own it: "If a man owns a little property, that property is him, it's part of him and it's like him. If he owns property only so he can walk on it and handle it and be sad when it isn't doing well, and feel fine when the rain falls on it, that property is him. . . . Even if he isn't successful he's big with his property" (*Grapes* 39). Such feelings descend directly from the dual myths of the frontier and the garden. The frontier myth posited that land in the West was not inhabited by anybody with legal rights and that the strength of the nation lay in its boundless and unsettled western frontier. The myth of the garden held that the land would yield bountiful harvests to any American willing to work it. Any failure in these natural laws was necessarily transitory and had no lasting relevance. Unfortunately, these precepts were disproved by the Okies' experiences in both Oklahoma and California. After a prolonged drought revealed the unsustainability of their farming methods and drove them from their homes, the wet-dry cycle in California nearly caused their demise.

Not only did meteorological laws conflict with their yeoman belief system, the Okies also found their way of life colliding with the policies of a nation committed to corporate capitalism. Empiricism and a bottom-line mentality created rigid parameters for decision making. For agrarians, land constituted a part of themselves and their culture, but banks and corporations translated it into assets on a balance sheet. Where the Okies spoke of "bad years," account managers acknowledged the reality of sparse rainfall and a semiarid climate. Historical climatic patterns decreed that "bad years" for rainfall were the norm for the Plains, a fact that made tenant farmers a poor investment. In addition, years of drought and overreliance on nutrient-draining cash crops devastated the land. Those keeping accounts looked to squeeze out every vestige of production before abandoning it for more lucrative investments:

> But you'll kill the land with cotton.
>
> We know. We've got to take cotton quick before the land dies. Then we'll sell the land. Lots of families in the East would like to own a piece of land. (*Grapes* 34)

For banks, it became a matter of short-term profit at any cost—extracting the land's limited value and then discarding it for more lucrative investments. But for the tenants, faceless corporate "monsters" intentionally destroying the land's fertility through overcultivation of cotton, a nutrient-draining crop, provoked violent, if futile resistance (*Grapes* 35). Paradoxically, however, the Joads and their neighbors had often planted cotton and, at the time the novel opens, are sharecropping frenziedly in order to build up a stake to take west: "The whole bunch of us chopped cotton, even Grampa" (*Grapes* 90). The differences between the Okies and the banks lay more in scale and philosophy than methodology and eventual result. Both sides participated in the capitalist mechanism, but the banks were better adapted to thrive within it.

Treating the nation's breadbasket as an expendable resource assumed an infinite reservoir from which to replace it. Short-term profiteering posits that the future will take care of itself. It depends on a telos of inexhaustible plenty, a concept central to the frontier and garden myths. These myths have shown remarkable durability: the onset of the Industrial Age and accompanying supremacy of corporate capitalism did not eradicate them, but simply adapted them to the twentieth century.

Richard Slotkin offers an intriguing explanation for this transition. He argues that the systems of myth and ideology that developed in the United States depended on a positive valuation of physical migration, which revolved around two geographical poles: the "metropolis" and the "frontier." The metropolis must have negative associations, or no one would want to leave, while the frontier has to offer riches enough to satisfy our dreams. Emigrants suffer in the wilderness while temporarily regressing to a more primitive state. The results, though, more than compensate for the ephemeral loss of civilization's comforts: "The completed American was therefore one who remade his fortune and his character by an emigration, a setting forth for newer and richer lands; by isolation and regression to a more primitive manner of life; and by establishing his political position" (*Environment* 35).

This description offers striking parallels to the Joads' saga. Slotkin's analysis takes the city, or the metropolis, as the emigrant's point of departure, but substituting the Dust Bowl region does not interfere with the argument. Since the trappings of the Industrial Revolution came late to the Plains, the region lacked the large, mechanized urban areas that posed such an effective antipode to the wilderness frontier. Instead, mechanization and factory farming—both consequences of industrialization—provided the major impetus that drove families like the Joads from the Plains. In the Dust Bowl, wage-slavery and the specter of starvation resulting from technological and economic displacement offered the negative contrast to the frontier. The Okies' choices, in Steinbeck's view, were either to serve the enemy by driving a tractor through their neighbors' homes and raping the land with machinery and cash crops, or to leave.

When the Joads first came to Sallisaw, Oklahoma, they endured isolation and primitive conditions while homesteading their land and seeking to fulfill yeoman ideals. Aridity and untenable agricultural practices caused the dream's collapse, forcing thousands of people like the Joads to again move west. This time they settled in California, geographical border of the once limitless frontier and now a privatized corporate fiefdom. Once more, the Okies suffered primitive, dehumanizing conditions while attempting to exercise their supposedly inalienable human rights. But the growers' cartel had disenfranchised them even before they arrived, forcing them into a nomadic existence designed to destroy the homesteading instinct so central to the frontier myth.

Despite uncountable acres lying fallow, no land was available for the Okies. Their dreams of subsistence farming proved fundamentally incompatible with a system designed to allow a select few to grow vastly wealthy on the toil of disenfranchised adherents to an old American Dream. What ultimately kills Jim Casy and exiles Tom Joad is—as in Slotkin's paradigm—an urgent desire to participate in the political process. They do not succeed because the growers' control over water rights allows them complete dominion over the local government and media as well. I will discuss this phenomenon at greater length shortly. Its relevance here stems from water's role in the third major cause (in addition to aridity and unsustainable agricultural practices) for the Okies' westward migration: inadequate irrigation and a perceived drought.

Steinbeck and Water

Steinbeck's humanistic bent impelled him to focus on the human dimension of the agricultural calamity that drove the Okies west. However, the underlying motivation for both the Okies' behavior and that of the agribusiness concerns can be analyzed in hydrological terms.

Drought did not cause the Dust Bowl. Rainfall in the Southwest in the 1930s fell well within historical norms; cycles of drought are more common than periods of heavy rain.[8] An accurate account of the region's troubles should focus on the depression and local agricultural mismanagement. The depression, though, did not seriously affect the Great Plains until the onset of the Dust Bowl in the early 1930s. If local farmers had continued planting and harvesting cash crops at the rate they had in previous decades, the Plains might have escaped the worst of the depression. Unfortunately, by the end of the 1920s, they had borrowed heavily and expanded their acreage to maximize yields. When the crops failed and the "black blizzards" came, the national plague of poverty and joblessness infected the Plains as well.

By the 1930s, Plains farmers had plowed under virtually all the region's grasslands. Without sod or other vegetation to hold the topsoil in place, the land became highly vulnerable to ecological disturbance. When the drought hit, the land had no natural defenses with which to anchor that topsoil. The resulting dust storms stripped the land bare. Steinbeck offers this lyrical description of one such storm:

> The air and the sky darkened and through them the sun shone redly, and there was a raw sting in the air. During the night the wind raced faster over the land, dug cunningly among the rootlets of the corn, and the corn fought the wind with its weakened leaves until the roots were freed by the prying wind and then each stalk settled wearily sideways toward the earth and pointed the direction of the wind. . . . Now the dust was evenly mixed with the air, an emulsion of dust and air. Houses were shut tight, and cloth wedged around doors and windows, but the dust came in . . . [and] settled like pollen on the chairs and tables, on the dishes. The people brushed it from their shoulders. Little lines of dust lay at their door sills. (*Grapes* 2–3)

Those famous storms did not savage the entire Plains. The most severe wind erosion took place primarily in eastern Colorado, western Kansas, and the Oklahoma and Texas panhandles.[9] Yet if even the most drought-afflicted region had retained its indigenous vegetation, the dust storms would have been much less powerful and their damage drastically reduced. Instead, profit-oriented agriculture and ecological ignorance turned a cyclical shortfall of water into a catastrophe.

High-yield, single-crop agriculture is a dubious ecological proposition even in humid regions, but in the Southwest it becomes disastrous. Barring a shortfall of precipitation, the average homestead usually proved adequate for subsistence farming. The region could not, however, sustain the rigors of a market-driven agriculture, a task that agriculture's metamorphosis into an industry soon demanded.

Steinbeck condemns what he saw as a dissolution of the values so cherished by the people who settled the region—connectedness to the land coupled with love and gratitude for its sustaining them. Such reverence became obsolete with the ascension of factory farming.

> The driver sat in his iron seat and he was proud of the straight lines he did not will, proud of the tractor he did not own or love, proud of the power he could not control. And when that crop grew, and was harvested, no man had crumbled a hot clod in his fingers and let the earth sift past his fingertips. No man had touched the seed, or lusted for the growth. Men ate what they had not raised, had no connection with the bread. The land bore under iron, and under iron gradually died. (*Grapes* 38)

Steinbeck mourned this change in values but offered no viable solutions. Even as they cursed the technology that drove them west, Steinbeck's characters traveled in cars bought through the trade of their mules and watched with sadness as tractors did the mules' work in a fraction of the time.

The Okies formed the pivot point for the western land's transition from earth mother to degraded resource. As the yeoman ideal gave way to the demands of a market economy, the Okies adapted their methods to meet the new reality. Even while clinging tenaciously to a preindustrial, agricultural vision, the Okies grudgingly acknowledged the new dominance of the capitalist shift and the technology that accompanied it.

Muley Graves, unable to relinquish his ties to the land, did not go with his family when they moved west. Rooted to the place where he was born, Muley's rage against the dual inequity of bad land and evil bankers is steeped in the populist tradition:

> 'Cause what'd they take when they tractored the folks off the lan'? What'd they get so their margin a profit was safe? . . . God knows the lan' ain't no good. Nobody been able to make a crop for years. But them sons-a-bitches at their desks, they just chopped folks in two. . . . Place where folks live is them folks. They ain't whole, out lonely on the road in a piled-up car. Them sons-a-bitches killed them. (*Grapes* 55)

For Muley, the link with the land still stained with his father's blood is stronger than his ties to wife and family. He cannot leave even as he acknowledges that he is a living anachronism. Sharing a meal with Tom Joad and Jim Casy of rabbits he snared while roaming through deserted farms, Muley repeatedly asks, "You fellas think I'm touched?" (*Grapes* 53, 54).[10]

Muley's protests held little weight with a population caught up in the fervor born of technological advance. It did not matter if the land was poor because human ingenuity would transform it. No longer need the land yield forth its bounty; it would instead be mined and harvested. Modern agriculture provided the means to merge Henry Adams's classic juxtaposition of the dynamo and the virgin.[11] Through this synthesis, the earth ceased to be a virgin and became a wife.

In *To a God Unknown,* Steinbeck openly acknowledges the sexual bond between men and the land. After Joseph, the story's protagonist, literally makes love to the earth, the narrator matter-of-factly notes, "For a moment, the land had been his wife" (11).[12] Later, as Joseph gradually merges his life force with the earth, his epiphany and death come with the realization that rain is the lifeblood of the land and that without a reintegration of humans and the land, both will die. In *Grapes,* written a decade after *To a God Unknown,* Steinbeck again acknowledges the sexual link—this time in the form of rape: "Behind the harrows, the long seeders—twelve curved iron penes erected in the foundry, orgasms set by gears, raping methodically, raping without passion" (37). The hopeful synthesis portrayed in *To a God Unknown* has given way to violent sexual assault.[13]

As industrialism began to dominate the West, the accompanying mind-set fit a unique niche in the American dream of rugged individualism and merit-based achievement. Myra Jehlen argues convincingly that the uniquely American bond with the land and nature makes Americans believe anything they choose to do is right and natural: "The settlers' implementation of the continent's permanent contours and conditions . . . vivifying the land from inside . . . places the emerging social structures beyond debate, in the realm of nature. Those who assist the emergence of those structures, moreover, wield the power of nature itself" (57). One way Americans cast the conquest of land within the current political climate involved classifying irrigation as a struggle between the forces of good and godless communists. In 1960, Robert Kerr, former governor of Oklahoma and head of the Senate's Select Committee on Water

Resources as well as cofounder of the Kerr-McGee Corporation, summed up a widely held philosophy of the arid lands when he rhetorically asked, "Can a pagan Communist nation . . . make more efficient use of soil and water resources than the most advanced and enlightened nation in the world? Can ruthless atheists mobilize and harness their treasures of God-given wealth to defeat and stifle freedom-loving peoples everywhere?" (323–24).

Westerners could restore Eden, but doing so involved "reclaiming" their place in paradise through diligence and industry. Men would finish what nature had begun. Eden, ideologues hastened to point out, was an irrigated garden. Adam fell, but Americans would stand tall. By the time of the Dust Bowl, the Bureau of Reclamation had already emerged as a potent force in the shaping of the western landscape, intending to fulfill Powell's credo of "rescuing" and "redeeming" the land from aridity.

The true meaning of the word *reclamation* lost all significance in the technological assault on the region's hydrology. The verb *to reclaim* implies prior ownership; those seeking to irrigate the desert could make no such claim. Nevertheless, they would do whatever needed to be done to get water to the land and restore it to its imagined, bountiful state.[14] Any water that ran into the sea without serving some agricultural purpose was a providential oversight correctable through human diligence.

Denying the hydrological realities of the Southwest permitted westerners some fanciful notions about American ingenuity. Henry Luce's *Time* trumpeted the rediscovered limitlessness that irrigation technology brought to the frontier: "Irrigation experts are now convinced that the rapidly growing U.S. can expand indefinitely within its present boundaries" (qtd. in Worster, *Rivers* 266). This quotation is pregnant with contradiction. Indefinite expansion within acknowledged boundaries is, of course, self-contradictory. Attributing this ability to accomplish the impossible to the calculations of irrigation experts beautifully underscores the incongruities within western water policy.

Western land barons relied on irrigation to accomplish the impossible and ignored or destroyed anyone or anything that might interfere with the attempt. The Joads and their contemporaries were ill equipped to cope with the growers' zeal. They clung to traditional yeoman values even while participating in the market economy that was rendering them obsolete. Caught between two worlds, they could not linger in Oklahoma and set out instead for the land where word of mouth from previous migrants combined with a propaganda campaign by the corporate growers to entice exodusters westward.

As they traversed the migrant highway, the Joads met many who, like themselves, had readily believed the leaflets spread by agents of the California growers.

> Why, I seen han'bills how they need folks to pick fruit, an' good wages. Why, jus' think how it's gonna be, under them shady trees a-pickin' fruit an' takin' a bite ever' once in a while. . . . An' with them good wages, maybe a fella can get hisself a piece a land an' work out for extra cash. Why, hell, in a couple a years I bet a fella could have a place of his own. (*Grapes* 160)

Those handbills stood in sharp contrast to the misery that gripped the Plains.[15] There, families faced wrenching poverty and slow starvation while signs in shop windows ironically proclaimed things like: "Great Bargains in Real Estate. Bring Your Own Container."

That the Plains could no longer sustain the yeoman ideal did not necessarily spell the death of the American Dream for a dispossessed people, barely literate and ready to jump at any hope of salvation. The California growers cartel, already profiting from a cycle of wage-slavery based on the exploitation of Mexican workers, believed that additional workers could only increase their profit margins.[16] When the supply of cheap immigrant labor was drastically cut back by immigration restrictions, the growers recruited Dust Bowl refugees instead, luring them with promises of a vast, temperate paradise wherein they might re-create the homesteads they had been forced to leave.

This new myth of the garden presented an even more seductive exterior than the Plains. It adapted the Jeffersonian ideal to a region where husbandry was allegedly secondary to the munificence of nature. Grampa, before becoming overwhelmed by his attachment to the land he cleared and raised his family on, fantasized about bathing in a washtub full of grapes where he would "scrooge aroun' an' let the juice run down my pants" (*Grapes* 100). But this vision of unchecked abundance was a calculated product of the growers' propaganda mills. California agriculture owed less to nature's bounty than to human engineering. The landscape was "a forced plant—the product of irrigation. [T]he great farm valleys were [once] wastelands and deserts into whose reclamation has gone untold human suffering" (McWilliams 5–6). Agribusiness consortia dangled visions of their own wealth and massive landholdings before the Okies in order to fuel their (the cartel's) hegemony. And the irony of that vision, as Steinbeck depicts it, is that the growers were as alienated from their land wealth as they forced the Okies to be: "And it came about that the owners no longer worked their farms. . . . [T]hey forgot the land, the smell and the feel of it, and remembered only that they owned it. . . . And the owners not only did not work the farms any more, many of them had never seen the farms they owned" (*Grapes* 257). In the landowners' view, alienation from the land was an ingredient for success and profit. They considered themselves businessmen, not farmers, who packaged a product for sale.[17]

The California growers had grown immensely wealthy and powerful as the result of an uneasy but mutually profitable alliance with the Bureau of Reclamation. They had managed to consolidate the dual definitions of *garden* into one highly profitable vision of production and wealth. No longer could *garden* signify either a region of natural, providential splendor or an area of human-created agrarian abundance; the Edenic garden propounded by Gilpin and his nineteenth-century allies was completely replaced by the Baconian vision of a human-engineered paradise achieved through work and intellect. Humans—specifically men—had invented the tools necessary to subjugate nature. Those tools had brought water to the desert via centrifugal pumping and, more importantly, through the diversion of rivers.

Shaping the perceived objectivity of science to fit the needs of western agriculture enabled an elite group to gain control over the dissemination of knowledge, leading to their dominion over the region's geography.[18] Literally overnight, worthless land became incredibly valuable through shady, often illicit dealings that brought subsidized water to the region. The men whose schemes created this technological garden stood to profit most from its enactment, and it was they who formed the powerful growers' cartel that enslaved the migrants. Those who controlled the water controlled the entire regional economy, and that domination bled into every other facet of life.

Californian agribusiness's command over nature and drive for ever greater wealth required large, transient labor forces who would work for very low pay. The growers had traditionally indentured immigrants and other disenfranchised groups since little public outcry arose from their mistreatment. The arrival of the Okies, a large, skilled, English-speaking labor force whose migrant status left them bereft of governmental protection, appeared to be a tremendous windfall. In the novel, however, the latent power of the oppressed becomes the looming threat to the water-based oligarchy.

The Okies come to embody Marx's concept of alienated labor.[19] Their corporate oppressors force them to work ever harder and faster in order to eke out a subsistence, yet each hour worked and each piece of fruit harvested brings them closer to unemployment and starvation. They must also compete against each other by underbidding fellow workers in a futile attempt to participate in an exclusionary economic system. The growers, meanwhile, must dehumanize the workers, degrading them as they do the land so that their acts of subjugation can be perpetrated on objects beneath contempt.[20]

Steinbeck's *In Dubious Battle* treats the worker-grower relationship in terms of class struggle. In *The Grapes of Wrath,* he elevates it to the realm of epistemology, viewing the schism between workers and land barons as symptomatic of the larger issue of human alienation from the earth and as a catalyst for the synthesis of humans and their surroundings into an all-encompassing organismic one: "Three hundred thousand, hungry and miserable; if ever they know themselves, the land will be theirs. . . . And the great owners, who had become through their holdings both more and less than men, ran to their destruction, and used every means that in the long run would destroy them" (*Grapes* 263). The combination of the growers' self-destructive behavior and the unendurable conditions imposed on the Okies will lead eventually (Steinbeck suggests) to a radical reordering of California's agricultural class structure.

The cycle of poverty imposed on the Okies in California contained a seasonal period of starvation during the rainy season. Water again, this time through superabundance, became the immediate threat to the Okies' survival. When Rose of Sharon goes into labor, the men outside struggle frantically to erect a dam to keep the boxcar shelters dry. Water, priceless commodity and building block of life, now endangers the birthing process and threatens to starve an entire class of people. Both attempts—the birth and the dam—prove unsuccessful. As the floodwaters force the Joads to flee, Uncle John is assigned the task of burying the stillborn child. Instead, he co-opts the water, using it and the dead child to spread his message of despair and defiance.

Setting the infant corpse adrift on the floodwaters, Uncle John tells it to: "Go down an' tell 'em. Go down in the street an' rot an' tell em that way. That's the way you can talk. . . . Go on down now an' lay in the street. Maybe they'll know then" (*Grapes* 494). Driven from Oklahoma, where widespread refusal to acknowledge water's scarcity resulted in an unsustainable way of life, the Okies found themselves in a new region where water's seasonal abundance and scarcity were integrated into a sophisticated capitalist infrastructure with water at its base. As a disenfranchised and powerless class, the migrants had no opportunity to gain control over water and consequently could not participate in the dominant discourse. Uncle John's act represents an ephemeral yet powerful appropriation of the preeminent unit of capital. He uses water to convey a message of worker defiance aimed at the heart of the power structure.

The dual hopes for the migrants, according to Steinbeck, are class alliance and worker control over the tools of domination. When Tom Joad takes over the task of organizing the Okies from the martyred Casy, the class struggle takes a symbolic step forward. When Uncle John symbolically seizes control over the waters that enslave his people and threaten their lives, he takes another step toward toppling the ruling class.

Shortly after Uncle John's act of defiance, Rose of Sharon's gift of her maternal milk to another starving Okie demonstrates that Tom's, Casy's, and John's acts will eventually show result. Sheltered from the water by a barn, itself a potent symbol of the yeoman agricultural ideal, Rose of Sharon's offering of her breast to a fellow migrant demonstrates a class cohesion that remains powerfully intact. While her stillborn infant rots in the town below, Rose of Sharon breast-feeds an old man whose advanced state of starvation has caused him to regress to a prelingual state. Her act and the old man's condition represent the succoring of the infant movement toward social change.

Each act, while primarily symbolic, is also genuinely subversive. In these small acts of defiance and hope, suggests Steinbeck, lie the restoration of traditional ties between people and between people and the land. Despite their socialization into a culture in which water is both hoarded and feared, the Okies have not completely acquiesced to their role in the factory-farm mechanism. They retain their dreams of an idyllic land where the family farm reigns supreme and water and land are distributed according to need and connectedness to the land rather than amassed corporate capital and political dominance.

In the final analysis, however, the migrant dream of resurgent family farms reclaiming their place as the preeminent agricultural ideal cannot work in the arid lands. Water reclamation projects, because of their expense and complexity, require the participation of an elite, educated class. The projects inevitably become political pawns. The family farmer, loyal to a subsistence ideology and unwilling to exploit the land past its carrying capacity, cannot compete with wealthy, powerful, corporate interests that have no similar ethical constraints. For this reason, the novel, though hopeful, does not offer any quantifiable hope.

Worster takes this lack of an attainable goal to be the novel's major failing. Decrying the system of land distribution without explicitly condemning the accompanying hydrological autocracy leads to the specious notion that simply putting the land in the hands of the migrants will solve the region's problems. In a section of *Rivers of Empire* entitled "The Grapes of Wealth," Worster argues that Steinbeck never explicitly draws attention to the "elaborate hydraulic apparatus" that enabled agriculture in California. Succulent fruit and luxuriant cotton seem to spring from nature's goodness rather than from "the contrivances of advanced water engineering and the social organization it has required" (229). Since Steinbeck failed to acknowledge the inherent oligarchic nature of irrigation-based societies, he creates the false impression that equitable land distribution and a classless society will return the region to ecological stability. Unfortunately, there are no historical precedents for this vision.

Furthermore, in Worster's view, returning the family farm to the arid region without altering the national capitalist infrastructure will, given the Plains example, cause devastating ecological harm.

Worster's critique raises the problematic issue of Steinbeck's unrepentant affection for the family farm but does not, as I mentioned earlier, address the powerful critique of hydraulic society implicit in the novel's structure. That Steinbeck used water throughout the novel as an absent signifier suggests that he was well aware of its power and complicity in the region's power hierarchy. When, at the novel's end, Steinbeck suddenly introduces water as a tangible presence and powerful symbolic force, it empowers the migrants by allowing them to demonstrate their class cohesion and latent strength.

Structuring the novel in this manner permitted Steinbeck to criticize the extant hydraulic society more effectively than he could through overt polemics. Indeed, the novel's reception, both locally and nationally, bears witness to its power. That power underscores the most crucial flaw in Worster's argument. If the novel caused both the government and the nation at large to reevaluate federal irrigation subsidies for corporate growers, it must have effectively criticized the inequity and corruption infusing California's water-appropriation methods. The Hearst-Chandler-Copley broadsheets pilloried the novel and its author throughout California. Only after a *Life* magazine exposé and Eleanor Roosevelt's endorsement of the book's veracity did the tide of public opinion begin to turn in Steinbeck's favor.[21]

The rage and furor from the agribusiness conglomerates and their allies arose because *The Grapes of Wrath* shook the very foundations of the water-based oligarchy. Worster himself acknowledges this:

> Up to the very end of the decade, both the Bureau [of Reclamation] and the Department of the Interior were placidly moving forward . . . avoiding any cause for alarm on the part of the growers in California. . . . What changed all of that undoubtedly was . . . *The Grapes of Wrath*. . . . Suddenly, it became rather difficult for a liberal government in Washington to give subsidized, unrestricted water to groups like the reactionary Associated Farmers, to underwrite their labor policies and their concentration of wealth. (*Rivers* 245)

It appeared that the fundamental principles underlying reclamation could no longer escape critical scrutiny and reform.

Nevertheless, despite a temporary surge in popular and governmental concern, neither the novel nor the reform movement it generated achieved any significant, lasting change in western water policy. Pork-barrel appropriations bills continued to subsidize corporate growers who continued to behave as if they lived in a technologically controlled Eden. The migrants' struggle faded into the background with the outbreak of World War II, when the nation turned to the West once again to fuel the American war machine.

The Okies benefited from the wartime surge in production, finding work in munitions factories and other war-related industries. Relieved, the growers turned once again to immigrant labor, a class of people they could be relatively certain of keeping disenfranchised and powerless. So the cycle of exploitation resumed after only a brief hiatus. Public interest in the issue peaked again two decades later in the 1960s, when Cesar Chavez briefly managed to organize the Migrant Farm Workers Union into an effective national lobby. Nevertheless, both the dam building and worker exploitation continued.[22]

Only in the 1990s, after a prolonged drought and numerous aborted attempts at reform, did the Californian agricultural machine begun to sputter. Years of drought and insupportable agriculture in an arid land are on the verge of accomplishing what neither Powell nor Steinbeck nor any individual person could accomplish alone: decanonization of the myth of the garden and its accompanying myth of the frontier. These myths, dominant since the birth of the nation, eventually ran headlong into the realities of a closed frontier and a finite hydrology. In 1992, Steven Goldstein, spokesman for former interior secretary Manuel Lujan, aptly summed up the situation when announcing the curtailment of further water subsidies to California growers, saying: "We recognize . . . what a hardship this will be. But we cannot make it rain."[23] The progression in consciousness that preceded Goldstein's statement was neither smooth nor uncomplicated. It involved a fundamental shift in how Americans viewed the land, a shift redolent with controversy and conflict.

4

Waging Water: Hydrology versus Mythology in *The Monkey Wrench Gang*

Whiskey is for drinkin': Water is for fightin'.
—Mark Twain

"The dam?"
"Yes sir."
"Not the dam."
"Yes sir, we have reason to think so."
"*Not* Glen Canyon *Dam!*"
"I know it sounds crazy. But that's what they're after."
—*The Monkey Wrench Gang*

Whereas Steinbeck's lingering nostalgia for the Jeffersonian yeoman ideal in *The Grapes of Wrath* arguably diverts his attention from the true nature of the southwestern agricultural and hydrological crisis, Edward Abbey focuses his energy in *The Monkey Wrench Gang* on documenting the ramifications of western water policy and vociferously advocating its demise. *The Monkey Wrench Gang* offers a prescription for grassroots action to restore beleaguered western lands and give the canyon country back its heart: the Colorado River. The Glen Canyon Dam, in Abbey's view, killed a living river and sounded the death knell for the entire Southwest.

The Monkey Wrench Gang was published in 1975 but, according to Abbey's epigraph, relies on historical events that occurred "just one year from today." It depicts a world where the myth of the garden has evolved into uncontrolled, government-subsidized growth-mania that has flooded much of the garden and poisoned the rest. Part of the novel's implicit irony stems from the forces of subversion having come full circle. In the Owens Valley during the early part of the century, residents dynamited dams as acts of defiance against the Los Angeles Department of Water and Power. The city's appropriation of water rights had destroyed the valley's once-rich agricultural infrastructure, leaving the region barren and sere. Later in the century, Steinbeck depicts starving Okies enjoined by agribusiness and absentee owners from planting subsistence crops on land purposely kept fallow. That land that they did plant was sown with high-yield, nutrient-draining cash crops that destroyed the soil's ability to support life. By Abbey's era, a lush, green desert was neither the ideal, nor commonly perceived as attainable. With agribusiness and other ill-conceived ecological practices causing the western United States to suffer a faster rate of desertification than sub-Saharan Africa, Abbey chose not to portray frustrated farmers unable to reap nature's bounty from a technologically controlled Eden. Instead he created heroes committed to destroying the apparatus that engineered the garden in the first place. Rather than commit sabotage in support of the agrarian ideal—as Owens Valley residents did and as the Okies conspired to do—Abbey's characters subvert that ideal through acts of "ecotage."[1] As in the Owens Valley, Abbey's characters look to dynamite a dam, but for vastly different motives. They hope to drastically reduce western agriculture by restoring the land to its naturally arid, "unproductive" state.

The opposition between "productive" in an extractive economy and "productive" in a steady-state economy is stark. In an extractive system, production means depleting nonrenewable resources in order to generate energy and goods. By contrast, a productive steady-state economy emphasizes renewable resources and expending no more energy than the system can replace. This range of meaning within the term *productive* underscores its malleability and the consequent power of economic systems to shape the discourse.

To Abbey, the Glen Canyon Dam embodies a cycle of *faux potlatch* carried on by development interests with the aid and complicity of the government. Rather than following the traditional potlatch scheme, involving large-scale accumulation of resources that are then given away or destroyed in a wholesale display and redistribution of wealth, practitioners of the southwestern faux potlatch amass huge amounts of the region's most precious resource (water) behind incredibly expensive dams and then waste that water in a manner calculated to maximize personal gain.

The aim is not to redistribute wealth, but to hoard it through manipulation of the nation's ideological biases and economic apparatus. Abbey underscores this in the following passage from *The Monkey Wrench Gang:* "Beneath the superstructure the dynamo purred on, murmuring the basic message: Power . . . profit . . . prestige . . . pleasure . . . profit . . . prestige . . . pleasure . . . power . . ." (234). The dam may store water and generate hydroelectricity, but its true purpose, in Abbey's view, is to enrich a select few.

An edifice designed to hoard wealth for the elite runs counter to the central tenets of potlatch. Among the Oweekano Kwakiutl tribe, for example, chiefs always died poor because they potlatched so frequently. "The idea of hoarding or accumulating wealth for any other purpose would have been regarded as abnormal, unthinkable, even shameful. . . . Paradoxically, to give away wealth was to be wealthy" (Cole and Chaikin 11–12). Throughout this chapter, my use of the term *potlatch* refers to the ceremonial expenditure of wealth practiced primarily by Native American communities in the Northwest until the early part of this century.[2]

Potlatch is a generic term referring to the feast and dance rituals of northwestern tribes involving large-scale giving of gifts by the host. The lavishness of the gathering and the extravagance of the attendant gifts awarded prestige to the host while obligating guests to reply in kind at a future potlatch. Often, a potlatch left a prominent member of the community all but destitute. This poverty further enhanced his status. "After the food is consumed, the dishes are given to those who ate from them. After the potlatch ends the host gives away the planks of his house and is said to have 'gone all the way.' The host is then left with nothing but his great prestige and renown to clothe him" (Rosman and Rubel 179–80). Wealth among these tribes was amassed for the sole purpose of giving it away. A host who potlatched knew that he compelled attendees to do the same. In this way, wealth never stayed static, but instead constantly circulated among communities.[3]

Within this complex marriage of prestige and poverty lay an economic system entirely foreign to capitalism. Rather than a spiraling economy dedicated to maximizing personal gain, the potlatch mandated periodic redistribution of resources among the community. Currency had no value unless it remained in circulation. Therefore, accumulated wealth in a potlatch economy was an oxymoron; possessions gained value only through their distribution. Not surprisingly, the encroaching ideology and economic system of European-descended settlers showed little tolerance for the potlatch. Missionaries and Canadian and American legislators all but eradicated it by the middle of the twentieth century.[4] In place of the potlatch rose a corrupted, water-oriented faux potlatch that was endemic rather than alien to capitalism and could not function outside its economic borders.

While maintaining a facade of profitless expenditure, faux potlatch actually greatly enriches its practitioners. For example, diversionary structures carry Colorado River water into the desert, where it is used to cultivate alfalfa and cotton, though nonarid regions of the country produce more than enough of these crops. In fact, producing cotton and alfalfa in the West via subsidized water and corporate tax incentives actually drives down world prices and increases the financial burden on farmers who do not receive such subsidies. However, the farmers who grow the alfalfa and cotton in the desert prosper. This type of agriculture increases water's scarcity while offering little tangible benefit to the majority of the citizenry. Demand for water, already in short supply, heightens, and the resulting scarcity boosts its exchange value.

In a traditional potlatch, the expenditures of the growers and the government would cause a widespread redistribution of wealth among all members of the community while enhancing their own status as givers. The givers would then start reaccumulating assets while waiting for others to potlatch, thereby continuing the circulation process that precluded the monopolization of wealth. However, since the faux potlatch is a closed system, the growers' expenditures benefit no one but themselves. Their waste acts as a trigger mechanism, causing the leviathan state to demonstrate that it retains the power to enrich and thereby placate its powerful citizenry.

The state expends also, but to little obvious gain. Its actions serve as preventative maintenance—rearguard actions aimed at maintaining political dominance. Only the growers benefit. The grower's behavior is, of course, calculated to achieve just such an aim. Even though the faux potlatch involves orgiastic expenditure and profligate waste, it offers no community gain or resource redistribution.

The term *faux potlatch* draws on Georges Bataille's description of the bourgeois appropriation and retooling of the potlatch ritual for purposes of power consolidation and capital gain.

> As a game, potlatch is the opposite of a principle of conservation. . . . In the market economy, the processes of exchange have an acquisitive sense. Fortunes are no longer placed on a gambling table. . . . It is only to the extent that stability is assured and can no longer be compromised by even considerable losses that these losses are submitted to the regime of unproductive expenditure. (122–23)

Allan Stoekl argues that, in Bataille's view, bourgeois methodology bears only a superficial resemblance to potlatch, just as potlatch itself only slightly resembles Thorstein Veblen's vision of "conspicuous consumption." To Bataille, conspicuous consumption does not embody hedonism or economic inefficiency but is rather a perversion of the human need to destroy. The ability to destroy fungible assets without causing oneself harm or suffering any diminution in power reaffirms one's place in the power structure (xvi). This redefining of "destruction" as an inherently profitable, risk-free enterprise forms the essence of southwestern hydraulic capitalism.

Abbey asks that we trade in the ecologically pernicious faux potlatch and myth of the garden for the environmentally sound, steady-state economy implicit in true potlatch. As Bataille illustrates through the "Icarian Complex," humanity accumulates in order to expend—we rise in order to fall. Our species seeks the base.[5] Risk and fall are inherent predicates of the true potlatch ritual. Bataille compares the human condition to a "deliriously formed ritual poker. But the players can never leave the game. . . . At no time does a fortune serve to shelter its owner from need. To the contrary, it functionally remains—as does the possessor—at the mercy of a need for limitless loss" (122–23). In Bataille's view, the risk of loss permeates and enriches the potlatch. Likewise for Abbey, risk greatly enhances life in the desert: "Because freedom, not safety, is the highest good" (*Gang* 26).

In Abbey's world, blowing up Glen Canyon Dam is the ultimate potlatch. Even as it represents the quintessential act of profligacy—an apparent squandering of the region's most precious resource, the dam's demolition is appositionally a supreme act of conservation. This paradox reflects what Bataille argues is not a conflict between the exalted and the base, reason and unreason, but their synthesis. Just as ingestion cannot exist without excretion, just as death and life coexist conterminously, so too is expenditure linked with accumulation. Any disruption of the natural order of things represents unsustainable "Icarian Revolt."[6]

In the true potlatch system, dispersion and destruction of wealth provide the only justification for its accumulation. Destroying Glen Canyon Dam would entail demolishing a massively expensive edifice and releasing millions of acre-feet of water for no discernible profit. However, the land as well as those residents who lack water rights would benefit enormously from the reallocation of the region's water wealth. In this sense, the dam's demise would constitute symbolic potlatch. It would also undermine the commodity value of water, emphasizing its utility instead. Use value would thereby replace exchange value as the dominant mode of valuation, and the riverine ecosystem would be restored.

In addition, just as potlatch restores balance by destroying and reallocating wealth, so too would destroying Glen Canyon Dam begin the process of reestablishing ecological equilibrium in the canyonlands. With ecosystemic balance restored, the cyclical process of accumulation and expenditure could begin again without any predetermined advantage for agribusiness.

Reclamation, as noted earlier, implies that the region once yielded abundant water and crops to all who asked as well as prior ownership and domain. This vision, steeped though it is in the myth of the garden, runs counter to history. On the other hand, *restore,* as used by Abbey in *The Monkey Wrench Gang* and other writings, has a referent in the recent past. It advocates the return of the Colorado River ecosystem as well as that of the arid lands in the river basin to their predam states, a steady-state ecological economy wherein the wealth of the river, captured behind a dam and forced to enrich a powerful minority, is liberated and redistributed to the land and people at large.

Unfortunately, the Colorado's ecosystem can no more return to its predam state than the American auto industry can return to the days before the arrival of Japanese cars. Rivers flow unceasingly; they epitomize constant change. There is no static past, frozen in time, to which they can return. Dams irrevocably altered the river and the land and the habits of those who live in their shadow. One can no more "restore" a dammed river than unfertilize an egg.

"Reclamation" represents the ritualized hoarding and sham expenditure practiced by agribusiness, with wealth remaining in the hands of a privileged few. Reclamation ideology, as we have seen in Austin and Steinbeck, exploits the potent but nevertheless specious myth that the Southwest will bloom into a new Eden under the stewardship of American expansionism. It promotes venture capitalism cloaked in a Jeffersonian yeoman, pioneer mask. The region's ecology and natural beauty are acceptable casualties in the conquest of nature.[7]

Abbey confronts the ecological ramifications of this cycle and attempts to explode the myth of the garden providing its ideological justification. His characters conduct a fictional struggle against what Worster calls the "hydraulic empire," a "hydraulic society of the West, increasingly a coercive, monolithic and hierarchical system, ruled by a power elite based on ownership of capital and expertise [with people] organized and induced to run, as the water in a canal does, in a straight line toward maximum yield, maximum profit" (*Rivers* 5). Ann Ronald argues that Abbey's vilification of dams and dam builders lampoons the ideology of superabundance that enslaves rivers to the forces of corporate greed, advocating an insurrection that will "dismantle the dynamo" (139). While urging his vision of an agrarian anarchist steady-state economy, Abbey also champions "inhumanism," a term coined by Robinson Jeffers to describe people who reject the notion that human endeavor is the central aim of life and that all other life is of inherently lesser value.[8]

Within both economic systems—faux potlatch and the agroanarchist steady state—water forms the privileged sign; its control determines the power structure. The machinations of corporate capitalism work to negate water's use value. Because profligacy implies vast resources at one's disposal, wasting without concern for loss demonstrates control. Hence, water barons flaunt their hegemony and wealth while simultaneously creating a perceived need for ever-increasing subsidy and water infusion. This cycle is "faux" potlatch because it redistributes only upward—it has no beneficiaries beyond those who called it.

Those who do benefit profit handsomely from the transference of hydraulic capital. Creating an artificial surplus of water-intensive crops like cotton and alfalfa increases the scarcity and hence the price of water. This expanding shortage causes the state to divert more water to the growers' use. Water's exchange value increases even as agribusiness enjoys larger hydraulic subsidies. Since water resources remain finite, a larger share to agribusiness means a smaller share for everything else (including the desert ecosystem). Furthermore, elevated exchange value means that those not enjoying government subsidies must pay more for their water.

The Monkey Wrench Gang represents a rebellion against the continued faux potlatch carried on by western water interests with the aid and complicity of the Bureau of Reclamation. The gang's mission to blow up Glen Canyon Dam is embellished by numerous guerrilla actions on behalf of the desert and its human and nonhuman inhabitants. They aim to subvert the myth of plenty that simultaneously drives the corporate capital machine and destroys the land and hydrology of the very garden it purports to create. In Abbey's world, "If so-called civilization can violate the land, as it did in Glen Canyon . . . then so-called violence can be used to stop it" (Ronald 195).

George Hayduke and company look to replace systemic land abuse with a system of land use more aligned with Powell's original prescription for the region. Abbey views the dam as the height of hydrological hypocrisy and an insult to Powell's vision (the fact that the reservoir behind the dam is named Lake Powell exacerbates the insult). *The Monkey Wrench Gang* portrays human victims of the faux-potlatch cycle banding together to restore an ecologically stable society in keeping with Abbey's philosophy of "agrarian anarchy." "Resist Much, Obey Little"; the epigram from Walt Whitman that begins the novel aptly encapsulates Abbey's proposed solution to the southwestern ecological crisis. He advocates removing the self-perpetuating cycle of wage/gain profligacy and replacing it with a system based on ecological continuity and community-based government (*One Life* 25–28).

Among the region's principal villains, in Abbey's view, are the corporate growers for whom the vestiges of the agrarian myth have themselves become exploitable commodities in the pursuit of tangible wealth (water, capital). Trading on the enculturated national reverence for the farmer (a perquisite that small farmers had themselves long exploited), agribusiness increases its control over the privileged sign. As early as 1955, Richard Hofstadter argued, "In reality, the rest of us support the farmer; for industrial and urban America, sentimentally and morally committed to the ideal of the family farm, has undertaken . . . to support more farm-owners on the farm than it really needs under modern agricultural technology. It is in part because of the persistence of our agrarian traditions that this concession to the farmers arouses less universal antagonism than do the efforts of other groups menaced by technological changes" (8) In the American West, massive agricultural subsidies (primarily in the form of artificially low water prices) support the beef cattle industry (alfalfa and hay comprise two of the region's primary crops). Yet despite huge expenditures of water, grain, and topsoil, the seven Colorado River basin states produce only 23 percent of the nation's cattle value. The expenditures continue due to the preeminence of the cattle and agroindustrial lobbying complex. The power of the western agricultural lobby has grown even as the number of farmers has dwindled. A well-run, vocal organization, the agriculture lobby has used the national affection for the historic role of the farmer to consolidate a great deal of power, money, land, and water in the hands of an elite.[9] Abbey descries this phenomenon, labeling western cattlemen "nothing more than welfare parasites." In an essay entitled "Free Speech: The Cowboy and His Cow," Abbey turns his barbed wit on the American penchant for romanticizing cowboys. He maintains that both cowboys and their profession are greatly overrated and that altogether too much honor is heaped on people who spend their lives "contemplating the hind end of a cow" (*One Life* 18).

Since growing alfalfa and cotton in the desert requires massive infusions of water, their cultivation in an arid region makes little ecological sense. The justification for the production of enormous quantities of these crops in the arid West must therefore be social rather than scientific. It arises from a sense of entitlement buttressed by technology. Wasting water in the desert while spending immense sums to "reclaim" the land from aridity theoretically demonstrates that even the ecologically impossible yields to American ingenuity. Nature became another tool for the ascendancy of the chosen people. In the words of Emerson, "Nature is thoroughly mediate. It is made to serve. It receives the dominion of man as meekly as the ass on which the savior rode" (*Works* 45).

Fulfilling the American destiny handed down from the Almighty required that humans make a green world out of an arid land. This ethos exemplifies the uniquely American combination of technological ascendancy and perceived divine right. God and technology theoretically would provide the means through which humanity would dominate and then achieve independence from nature.[10] The settlement of the West owed much to this conjoining of human industry with the presumed will of the Deity. It allowed humans to co-opt God's will and turn it into human destiny and then use that notion of human destiny to expand the fortunes of a select few. The state would subsidize this process.

By 1975, when *The Monkey Wrench Gang* was published, the ideological mechanism underlying a self-perpetuating cycle of expenditure and subsidy had long been in place. Those who would waste water had learned to turn to the Bureau of Reclamation and cry poverty. They found a sympathetic and powerful audience. The bureau responded by building dams and diversions that provided water to raise more crops and cattle, which in turn required more water, ad infinitum.[11]

Bernard DeVoto expressed the western attitude toward the federal government as, "Get out and give us more money." The phrase reflects the schism between the pioneer myth whose accompanying animus toward external authority continues to influence western ideology and the realities of the Reclamation era. None of the monumental reclamation projects that pepper the West could have been built without federal subsidy. Meanwhile, very few of the enormous ranches and farms that supposedly embody the western way of life could exist without these reclamation projects. This leads to an ironic counterpoint wherein westerners fiercely assert their independence while loudly demanding increased federal aid.

In essence, the powerful members of the community ritually expend vast amounts of water in a way that vaguely recalls the American tradition of yeoman husbandry and infinite ambition. While their methods and ideology bear little resemblance to the Jeffersonian notion of family farms and subsistence agriculture, they do purport to subdue the forces of nature to the needs of humans and thus, in some sense, embody the dual notions of American ingenuity and manifest destiny. Their propaganda seized on this shared goal and successfully conflated agribusiness and the mythic western figure. Both claimed as their symbol a man who is independent, one with the land, and dedicated to the perpetuation of the "American" way of life. This yeoman facade struck and continues to strike a resonant chord within the government, engendering an almost Pavlovian political response.

The state, that is, the Bureau of Reclamation, affirms the national ideology and its own entitlement to govern by helping corporate growers waste water under the guise of taming nature. The state also participates in the cycle through massive federal spending and wasteful hydroexpenditures of its own. The most obvious and arguably the most ecologically repugnant of these expenditures are the huge federal dams festooning the West.[12]

Glen Canyon Dam, object of the monkeywrenchers' animus, went on line in 1964 at a final cost of $300 million.[13] It had a 27 million acre-foot storage capacity at the time of its completion, though it annually forfeits 70,000 acre-feet of storage space to siltation. It also suffers annual evaporation losses of 450,000 acre-feet.

Meanwhile, the Colorado River is vastly oversubscribed. Even without calculating the massive evaporation losses of Glen Canyon and the many other reservoirs constructed along its length, the Colorado's annual flow almost never reaches the figure ascribed to it by the various states, courts, and federal agencies when they divided its water among the basin states and Mexico. For example, the Bureau of Reclamation originally apportioned 16.8 million acre-feet of Colorado River water despite the river's average flow of 13.9 million acre-feet. This led to the Colorado being called a "deficit river," meaning that more water was appropriated from the river than it actually had. Labeling the river as deficient permitted the river's managers to blame the water for not being sufficiently plentiful. It allowed people in the West to continue behaving "as if the river, not the plans and works of those who used its waters, were to blame for its shortcomings" (Fradkin 15).

Those shortcomings have become glaringly obvious. It has been over a generation since the Colorado reached its mouth in the Gulf of California. When the little water that remains after dams and diversions reaches the delta in Mexico, it is three times as saline as ocean water and flows so weakly that it disappears into the sand long before reaching the Gulf.

Acknowledging the impossibility of attaining hydrological equilibrium under the current scheme requires accepting limitations on growth. And that concept, until very recently, remained something that neither the government nor regional agrointerests could entertain. Building dams affirmed—at least for the short term—state dominance over water and the accompanying right of its distribution. Dams also created reservoirs, which generated the illusion of an ample supply of water. But the very acts aimed at demonstrating control and abundance instead revealed a critical shortage. And in times of shortage, the rich generally become richer while the poor grow poorer still.

Such are the consequences of the agrarian myth in the Southwest. Over the years, both sides of the debate regarding development in the arid regions have co-opted Powell's doctrine for their own ideological gain. Whereas Abbey quoted Powell to commemorate Abbey's last trip through Glen Canyon (*Desert* 192), the Bureau of Reclamation attaches Powell's name to the reservoir that flooded the canyon and published a book entitled *Lake Powell, Jewel of the Colorado.* Abbey is all too aware of this irony and excoriates those responsible: "The impounded waters form an artificial lake named Powell, supposedly to honor but actually to dishonor the memory, spirit and vision of Major John Wesley Powell. . . . Where he and his brave men once lined the rapids and glided through the silent canyons . . . the motorboats now whine, scumming the water with cigarette butts, beer cans and oil, dragging the water skiers on their endless rounds, clockwise" (*Desert* 173–74). Viewing the dam and reservoir as blights on the landscape and as a desecration of the canyon provides an easy justification to the plan to destroy them.

In *The Monkey Wrench Gang,* the people Abbey places on the front lines of the struggle against the despoliation of the western landscape have varying motivations for their shared ideals. Doc Sarvis and Bonnie are indignant ecovandals. Doc destroys road signs because they interfere with his serenity, while Bonnie, in the beginning, acts mostly out of companionship for Doc. She later develops her own sense of outrage, especially when confronted with the ravaged landscape of a mountain clear-cut (*Gang* 209–12).[14] Doc participates reluctantly and with abject pacifism. He is the humanist of the gang, acting out of a love for humanity as well as for the landscape. His actions manifest a quixotic desire to resist the fatalism that otherwise infuses his outlook.

> "The reason there are so many people on the river these days is because there are too many people everywhere else. . . . The wilderness once offered men a plausible way of life," the doctor said. "Now it functions as a psychiatric refuge. Soon there will be no wilderness. . . . Soon there will be no place to go. Then the madness will be universal. . . . And the universe goes mad." (*Gang* 60)

Rebelling against what he sees as inevitable, yet also dedicated to serving the very people responsible, Doc lumbers gregariously into the breach. At tale's end, he and Bonnie even sacrifice their freedom to go to the aid of Bishop Love, their chief antagonist.

Nevertheless, Doc's role in the gang's ecotage should not be underestimated. As resident sage, his imprimatur lends dignity to their exploits, and his checkbook enables them. Despite deriving his wealth from the technologically enhanced ability of medicine to treat the needs of an expanding population, Doc good-naturedly funds a guerrilla war against the source of his creature comforts. He is in effect redistributing his wealth in a manner that will, if successful, force widespread emulation. The novel's conclusion validates the depth of Doc's commitment as he gives up his lucrative practice to tend to a small, remote community.

Seldom Seen Smith and George Hayduke represent homegrown victims of the technocracy. Smith, a river guide, has seen his livelihood defaced and his place of business drowned. His official address remains Hite, Utah, though that town now lies beneath tons of silt and water. Slow to anger but deceptively fanatical, Smith is the one who suggests destroying the dam, a telos that soon infects the gang's entire existence. His quest is quasi-religious:

> "Dear old God," he prayed, "you know and I know what it was like here before them bastards from Washington moved in and ruined it all. You remember the river, how fat and golden it was . . . Remem-

> ber the deer on the sandbars . . . and the catfish so big and tasty and how they'd bite on spoiled salami? . . . Remember the cataracts in Forty-Mile Canyon? Well, they flooded out about half of them too. . . . Listen . . . There's something you can do for me, God. How about a little old *pre*-cision earthquake right under this dam? . . . Okay, God, I see you don't want to do it just now. Well all right, suit yourself, you're the boss, but we ain't got a hell of a lot of time. Make it pretty soon, goddammit. A-men." (*Gang* 32–33)

Smith cannot decide whether to pray or to chastise God. As a Mormon, he is obligated by church dogma to have as many children as he can (Bishop Love, his nemesis, has eleven).[15] He is polygamous, a lifestyle that runs counter to modern church teachings and that only the most traditional Mormons continue to embrace. Smith, though, is childless, a renegade "jack" Mormon. He is uncomfortable with the teachings of his church but cannot renounce them entirely. Even his name, "Joseph 'Seldom Seen' Smith" suggests an inherent ambivalence about Mormonism. Though a namesake of the church's founder, Smith's nickname suggests an incongruity between his lifestyle and the tenets of the Latter-day Saints.

In a sense, Smith's theological quandary mirrors the dilemma of the expanding Southwest. As a river guide, he makes his living taking tourists into the canyonlands. His welfare thus depends on a constant influx of people into the region. Despite this apparent need for population expansion, Smith makes it as difficult as he can for people to visit or settle in the Southwest. He blows up bridges, destroys surveying equipment and power stations, and commits various other acts of ecotage. His dilemma resembles Doc's, and both men display similar reluctance at the start of every operation. The only anchor for Smith's worldview is the need to destroy Glen Canyon Dam:

> The blue death, Smith called it [Lake Powell]. Like Hayduke, his heart was full of a healthy hatred. . . . He remembered the golden river flowing to the sea. He remembered canyons called Hidden Passage and Salvation and Last Chance and . . . some that never had a name. . . . All of these things now lay beneath the dead water of the reservoir, slowly disappearing under layers of descending silt. How could he forget? He had seen too much. (*Gang* 31)

The theological overtones of the canyons chosen are no accident. Smith's religious association with the dam's demise recurs throughout the novel. Monkeywrenching provides him with a kind of theological walkabout—a quest for the presence of a righteous God. He believes in right and wrong but remains uncertain as to the existence of a just and merciful God:

> "All we need is to make one little crack in it, Doc. One crack in that dam and nature, she'll take care of the rest. Nature and God."
> "Whose side is God on?"
> "That's something I aim to find out." (*Gang* 258)

While Smith's fervor remains low-key and resonates with his own brand of theology, Hayduke evinces the hard-core fanaticism that drives the gang.

> Hayduke thought. Finally the idea arrived. He said, "My job is to save the fucking wilderness. I don't know anything else worth saving." (*Gang* 211)

As a Vietnam veteran, former POW and Green Beret medic, Hayduke brings his own brand of post-traumatic stress disorder to the war to save the West.

> I get a pension too. Twenty-five percent disability. Head case. One quarter lunatic. I must have a dozen checks waiting for me at the old man's place. . . . They really wanted to court-martial me but Mom wouldn't stand for it. Anyhow, when I finally got free of those jail-hospitals and found out they were trying to do the same thing to the West that they did to that little country over there, I got mad all over again. (*Gang* 330)

Hayduke often finds himself flashing back to his days in the jungle fleeing the helicopters that were supposedly on his side. A former POW half-brainwashed into a Vietcong, George Washington Hayduke lived in terror of American forces. His return to the States offers little respite. He is again pursued (the novel's structure suggests) by those same forces that drove the United States into an unwinnable, resource-draining war.

Hayduke is a caricature—a monomaniacal force of nature with a taste for Schlitz. His brute determination and survivalist tactics, coupled with an almost suicidal bent, enable him and his cohorts to successfully carry out much grander operations than they otherwise could. Even his colleagues worry about his sanity and then, because of their association with him, question their own.

Because of his quixotic drive, it seems only right that Hayduke should meet and bond with Jack Burns, the mythic Quixote-figure of several other Abbey novels. Burns is the "Brave Cowboy" of the novel by the same name, an anachronism who refuses to die and whose lifestyle and sense of loyalty have no place in a West dedicated to personal gain and terrestrial degradation. In *The Monkey Wrench Gang,* Burns is transformed into an ecowarrior—a vagabond like Hayduke except that he rides a horse instead of driving a jeep. Together they symbolize the past and future of the West.

Burns has come from the past to help Hayduke re-create a world where independence and a land-based value system will once again hold sway.[16] Burns also represents Hayduke's lone-wolf trait carried to an extreme. As a mythic figure who comes out of the night and refuses to give his name (until the novel's sequel, *Hayduke Lives*), Burns symbolizes what Hayduke will soon become—a ghostlike figure, more legend than man. Interestingly, Burns also represents what even Hayduke knows is the dangerous world of solipsism and make-believe: "For even Hayduke sensed, when he faced the thing directly, that the total loner would go insane. Was insane. Somewhere in the depths of solitude, beyond wildness and freedom, lay the trap of madness" (*Gang* 106). Abbey extricates Hayduke and Burns from their seemingly inevitable madness by bonding the two together at book's end. As their existence fades into legend and society assumes them dead (Burns by the side of the road in *The Brave Cowboy* and Hayduke in a fusillade of bullets at the end of *The Monkey Wrench Gang*), they somehow find each other and that bonding saves them from the otherwise inevitable madness.

Abbey does not dispute Hayduke's instability, but he does suggest that Hayduke is a product of his time and that the status quo also contains a form of madness that will require a certain amount of fanaticism to dislodge it: "Viewing it this way, we can see that the religion of endless growth—like any religion based on blind faith rather than reason—is a kind of mania . . . indeed a disease. . . . Growth for the sake of growth is the ideology of the cancer cell. Cancer has no purpose but growth; but it does have another result—the death of the host" (*One Life* 21). Throughout his oeuvre, Abbey advocates desperate measures to contravene the rampant destruction that the corporate-state monolith hungry for water and land wreaks on the landscape.

In prior times, writers of the West created scenarios where characters drew their strength from their unending battles to carve a life from the majestic yet merciless land.[17] Abbey turns this formulation around. The land's survival depends, in the case of Hayduke, on his resourcefulness and his power to inspire others to restore the ecosystem's balance. In *Desert Solitaire,* Abbey draws an important distinction between civilization and culture, calling civilization "the vital force in human history" and culture "that inert mass of institutions and organizations which . . . drag[s] down the advance of life." Civilization, he claims, is "the wild river" and culture "592,000 tons of cement" (277).[18] It is up to Hayduke to rescue the West from culture. Hayduke embodies Abbey's oft-stated desire to replace our current, highly integrated social system with a loose knit series of autonomous societies living in harmony with their surroundings.[19] While Hayduke himself is not so thoughtful, his goal of "sav[ing] the fucking wilderness" and his blows against the faux-potlatch system breathe behavioral life into Abbey's polemical ideas.

In *Abbey's Road,* Abbey declares, "If we can draw the line against the industrial machine in America, and make it hold, then perhaps . . . we can gradually force industrialism underground, where it belongs, and restore to all citizens of our nation their rightful heritage of breathable air, drinkable water, open space, family-farm agriculture, a truly democratic political economy" (137). This argument recalls Bataille's recipe for the dissolution of the bourgeois state. Abbey contends that industrialism forms an autocratic monolith that usurps the basic human rights and needs of the world's inhabitants. Potlatch (though Abbey does not use the term), as a socially encoded redistribution of ecological wealth (water, air, open space, etc.), must replace the faux potlatch practiced by the industrial oligarchy, or the biotic economy will implode.[20]

According to Bataille (and Abbey), industrialism's capacity to cause vast ecosystemic harm stems from the ascension of science to the seat of highest authority within the technoindustrial paradigm: "It is possible in all freedom to be a plaything of evil if evil itself does not have to answer before God" (Bataille 49).[21] Science and the machine displace natural theology, allowing technology to remake the earth in the image of human ingenuity. In the United States "Americans saw themselves as building their civilization out of nature itself, as neither the analogue nor the translation of Natural Law but its direct expression" (Jehlen 3). Rather than interpreting Original Sin as the necessity to remake human behavioral norms, it signaled the need to remake the earth through scientific advance into a facsimile of the original Eden.

In the Southwest, this renovation revolved by necessity around the manipulation of water and its diversion for use in irrigation, hydroelectricity, and urban sprawl. For Bataille, the myth of the garden may be seen as Icarian revolt—a denial of the cyclical rise and fall of all things. He acknowledges Nietzsche's notion of the simultaneous ascension-martyrdom of Christ as the ultimate expression of *ressentiment* power and imputes similar motivations to his own formulation of "imperative heterogeneity"—a pious materialism obviating all impulsions but the pursuit of power. Attempting to remake Eden in our own/God's image denotes a continued refusal to recognize the cycle of rise and fall and the ascension of the base (as represented by Hayduke as well as by the desert itself) inherent in the health of the economic system as well as the ecosystem.[22]

Bataille sees no easy solution to this cyclic hoarding of power. Any attempt at subversion must border on the fanatical if it is to succeed. His description of such a fanatic recalls Abbey's Hayduke: "[T]he introduction of a lawless intellectual . . . into the world of legitimate thought defines itself at the outset as the most arduous and audacious operation. And it is evident that if it were not practiced without equivocation, with a resolution and a rigor rarely attained in other cases, it would be the most vain operation" (80). Obviously, Hayduke is not an intellectual. Rather than injecting a renegade theoretician into an extant philosophical framework, Abbey inserted an action-oriented individual into a functioning state apparatus.

The results of the gang's efforts are hopeful but inconclusive. Initially, things go well as the gang successfully carries out a number of monkeywrench operations. Eventually, however, they are tracked down and cornered by Bishop Love and the Search and Rescue Team. During a standoff in a box canyon, Bishop Love suffers a heart attack and Doc and Bonnie go to his aid and their arrest. Smith, half-starved and unable to run any farther, is captured in a public campground, while Hayduke appears to get shot to pieces and fall five hundred feet to his death.

Slowly, the outlook improves. Doc and Bonnie's humanitarian acts lead to the partial conversion of Bishop Love (a conversion he repudiates in the novel's sequel). The surviving three members of the gang stand trial for, and publicly repent of, their actions. Doc even converts to Mormonism. They retire from monkeywrenching and become law-abiding members of society living under the watchful eye of their parole officer. Glen Canyon Dam remains intact, and the faux-potlatch cycle seems destined to continue.

Into this supposedly peaceful status quo steps Hayduke, unrepentant and more dangerous than ever. After staging his death in the canyon, he goes into hiding and then reemerges having metamorphosed into a pure force of nature. No longer is he distracted by erotic need (for Bonnie) or even the standard appurtenances of twentieth-century existence. He has become a legend (Rudolf the Red) and an untraceable warrior for the canyonlands. Having landed a job in the bowels of the hated dam, he has every intention of fulfilling the gang's mission.

Furthermore, as the prolegomenon of the novel suggests (part of which is quoted at the beginning of this chapter), the movement has gained adherents, becoming larger than the actors themselves. So despite their forced diaspora, the gang's work continues even as its members fulfill their parole. Abbey suggests that though the time may not be right for the demise of faux potlatch, seeds have been planted that will eventually choke the hydraulically illegitimate state apparatus. Corporate control over the privileged sign is no longer assured. The forces of ecological stability will use whatever means are necessary to restore hydrological equilibrium to "the heart of the heart of the canyonlands."

> Somewhere under the heavy burden of water going nowhere, under the silence, the old rocks of the river channel waited for the promised resurrection. Promised by whom? Promised by Capt. Joseph "Seldom Seen" Smith; by Sgt. George Washington Hayduke; by Doc Sarvis and Ms. Bonnie Abzug, that's whom. (*Gang* 112–13)

In an odd postscript wherein life imitates art, EarthFirst!, a radical environmental group that advocates ecotage and whose symbol is the monkey wrench, took up the gang's mission as their own. In 1981, they unfurled a huge canvas crack on the face of the Glen Canyon Dam, announcing to all their goal, its total demise. Abbey, for his part, wrote the introduction to *Ecodefense,* the manual of ecotage drafted by Dave Foreman, the founder of EarthFirst!. Abbey also incorporated Earth-First! and other radical environmental actors into *Hayduke Lives,* the posthumously published sequel to *The Monkey Wrench Gang.* Having created a new myth of the ecowarrior that, like the myth of the garden, spawned real-life progeny, Abbey weaves a new story containing characters from fact and fiction. In doing so he simultaneously acknowledges both the potency of myth and the irony of its predominance.

5

The Environmental System: *Animal Dreams* and the Rhetoric of Environment

A river is more than an amenity; it is a treasure.
—Oliver Wendell Holmes

At the beginning of the twentieth century, Mary Austin believed in spite of herself that Reclamation would work to the good of both humans and the land. In the 1930s, John Steinbeck portrayed Reclamation's horrific social and ecological effects while still hoping that it could be retailored to benefit yeoman farmers. The 1970s find Edward Abbey raging against the desecration of the canyonlands and advocating the forcible destruction of Reclamation and all its trappings. By 1990, Barbara Kingsolver seeks to articulate a vision of a post-Reclamation American West. The problem she faces also plagued her forebears; it is extraordinarily difficult to articulate a vision for change in a language designed to resist it.

In *Animal Dreams,* Kingsolver attempts to describe a new western reality, predicated on an organicist, ecofeminist vision. She does not succeed. That does not mean, however, that the novel is a failure. Read as an ambitious effort to rethink social and ecological mores, *Animal Dreams* gains richness, offering a disturbing glimpse at the difficulties inherent in the construction of a post-Reclamation era.

Many of those difficulties are communication-based, and for that reason my analysis of the novel adopts a systems-theoretical approach. Systems theory provides a means for studying communication within complex systems. A system is an organization of components functioning as a unit to perpetuate the survival of the whole. A human, for example, is a biological system comprised of many function systems (digestive system, nervous system, etc.). Humans are themselves components of a larger social system, which in turn forms part of an ecosystem, and so on.

Systems are formed and maintained through communication among components. That communication constructs meaning. Meaning is not *transferred* through communication; it is *created* through communication. There must be mutual comprehension by both parties before meaning becomes realized. Without mutual comprehension, attempts at communication create only noise. If, for example, I am training a dog to sit, I may have to repeat the command over and over. The dog may roll over, heel, shake hands, or lie down in an attempt to find meaning in my words. Eventually, it will glean that I wish it to sit and at that moment—when the dog and I understand each other—my command ceases to be noise and becomes communication.

Animal Dreams tackles a problem far more complicated than dog training. The concepts Kingsolver wishes to communicate (and therethrough give meaning) have not been articulated before. There are no words to describe a post-Reclamation society. In fact, the language Kingsolver must use resists such ideas because, if actualized, the ideas will topple the current power structure. Kingsolver thus faces the dual challenge of formulating a coherent alternative to Reclamation using language that hinders her at every turn, and then communicating that vision to her readers. This chapter analyzes the efficacy of her attempt and places it in the context of the late Reclamation era.

Literature has historically provided a potent medium through which to stretch and prod language. Layers of meaning and the elastic divisions between fiction and nonfiction as well as among other literary genres have probed the form and function of text as text and as discourse. Environmental literature has done this out of necessity, since the amorphousness of both the genre and the term *(environment)* means that their respective meanings require constant reworking.

This lack of consensus triggers repercussions, especially with regard to the environment. Though most people can agree on the need for environmental reform in the West, there is much less agreement as to the meanings of *environment* or *reform*. Consequently, communication stalls and attempts at "environmental protection" break down. In this way, linguistic inertia protects the status quo. This centrality of communication to social change combined with the lack of a static definition of environment forms an imposing barrier to an ecologically viable society.

We have seen that the etiology of Reclamation lay in a national allegiance to the American Dream, not in an empirical, science-based cosmology. Restoration dogma, meanwhile, mirrors a growing disenchantment with the ecological destruction wrought by an extraction-based social system, and is an effort to address the conservation movement's inability to craft alternatives. Yet any alternatives to the present system, including Restoration, can only come about through language, and language poses a formidable obstacle indeed.

Reclassifying Resources

When parties communicate, each party represents an external perturbation to the other. For example, suppose I engage you in conversation as you lie lost in reverie, dreaming of the previous evening's romance. You wish to return to your thoughts and, so, gently disengage yourself from further conversation. You use language (both verbal and body) to facilitate your return to equilibrium. I disturbed you (an external perturbation), and you responded through language in a manner designed to eliminate the disturbance.

Whenever two or more parties communicate, that interaction represents a departure from the status quo. At times, the discourse may describe a perturbation of great magnitude (perhaps a war requiring delicate diplomacy to subdue) or a pleasurable situation that it would be beneficial to prolong (such as the previous evening's romance). In both instances, language provides the mediating force that enables coexistence. I will discuss this facet of language in more detail shortly. For now, it suffices to realize that language facilitates homeostasis while discouraging radical change. This seems only logical; language is an invention of the system it maintains and is shaped by those in power as a means of retaining dominance. It has evolved as a tool for maintaining the status quo.

Discourse is possible because function systems share the goal of the larger system's autopoiesis, the process by which systems maintain and reproduce themselves.[1] That common goal of all the system components creates a shared context through which to construct meaning. In the human body, the continued survival of the particular human provides the common context. Like all complex systems, humans contain many function systems for handling particular facets of their autopoiesis. The nervous, circulatory, and digestive systems fulfill different functions dedicated to the maintenance and continued viability of each human being. The same diversification and delegation takes place in social systems dedicated to the reproduction and maintenance of society. The legal, political, educational, and economic systems function independently, while remaining dedicated to the good of the whole.

Discord within a system becomes inevitable as complexity increases. Function systems can have distinct, competing goals even while sharing the larger imperative of maintaining the unity. Teachers and parents may differ on how best to educate a child, two developers may disagree on how best to build on a tract of land, and so on. While the disputants disagree, they share a common medium through which to mediate their conflict. Without that common context, communication devolves into noise.[2]

In a social system, a conundrum arises when people perceive the system's operation as flawed yet cannot conceive of the means to correct that flaw. This is a phenomenon distinct from the self-regulatory activities of different function systems, as when the judicial and political branches differ over the interpretation of a particular statute. Radical reform would involve instigating fundamental changes in the structure and boundaries of a system. Language offers the only means through which to effect these changes, even as its mission constrains its vocabulary to prevent such radical shifts.[3] As a result, system opponents must work within a medium designed to subvert their agenda. This paradox has serious implications for those wishing to depose the Reclamation era and install a more (hydro)logical regime in the West.

The prevailing definition of environment was produced by a system that views its surroundings through a lens dedicated to economic expansion. It evaluates all things according to their potential value, which then determines their use. Even before technology reduces its matter and energy to products, Timothy Luke observes, "Nature already is transformed discursively into 'natural resources'" ("Environmentality" 58).

The system's validation and perpetuation lie in linguistically configuring its surroundings to reflect its aims.

Animal Dreams underscores the flexibility of the term *resource.* Classifying an object as a resource confers value. A natural resource conjures images of an asset ripe for exploitation. Designation as a resource removes an object from the shadowy realm of "environment" and tosses it into the swift-flowing stream of transactive capitalism, where its value becomes a product of discursive maneuvering. In Kingsolver's fictional town of Grace, Arizona, land and water are vital resources for the disenfranchised indigenous community. By contrast, the privileged members of the economic system (the Black Mountain Mining Company) measure the region's resource potential by its subterranean ore content. Once the ore is extracted, the land becomes permanently devalued.

Because water is ancillary to the mining process, the company gauges water's worth by measuring its exchange value in a market economy, a value that increases with scarcity. For the inhabitants of Grace, however, scarce water means less food. Consequently, for the townspeople, water's worth appreciates in direct relation to availability and use value. Though they would pay more for water during times of scarcity, that is a function of exchange value, not a measure of water's functional worth in a subsistence economy. The differing methods of valuation of the mining company and the Grace townsfolk stem from divergent notions of environment and are fundamentally incompatible.

In systems-theoretical terms, *environment* refers to everything that is *not* the system. It includes everything with which the system cannot communicate. Yet the environment is not merely chaos lurking outside the system's boundaries; it is itself a network of systems for which our system comprises *its* environment. The two sides, system and environment, share a dynamic border that shifts and flows in response to perturbation. A system responds and adapts to environmental perturbation in a manner designed to ensure its survival (autopoiesis). If the system survives, it will have gained complexity, enabling it to better cope with similar perturbations in the future. In other words, it will have evolved.[4]

The environment (i.e., whatever lies outside a system's boundaries) cannot be articulated because it lies, by definition, outside the realm of communication. As the sum total of everything not the system, the environment is necessarily far more complex than the system itself. Meaning lies in the relationship between the system's components, while the system's stability depends on its ability to reproduce and function despite and because of ongoing environmental disturbance. *Animal Dreams* represents an attempt to replace the dominant system (and discourse) with one that privileges a steady-state economic apparatus.

Kingsolver advocates replacing atomized, patriarchal science with an organicist worldview predicated on ecofeminism and Native American ecospirituality. In this new reality, land-based values would enjoy a higher status expressed through a language designed to perpetuate those values.[5] Seventy-three years earlier, Mary Austin encountered problems when attempting to manipulate the newly formed Reclamation rhetoric to fit a nascent environmental agenda. By 1990, nine decades of linguistic evolution allow Kingsolver more flexibility, though the obstacles remain enormous. A closer examination of *Animal Dreams* reveals the scope and nature of many of these barriers.

Systems Theory and *Animal Dreams*

Codi Noline, the novel's protagonist, has no link with her past. Her traumatic childhood in Grace—characterized by her mother's death; emotional neglect by her father, Doc Homer; and her own miscarriage or abortion at age fifteen—leaves her disconnected from the passionate environmentalism of her youth, and prone to listlessness. Though smart and capable, Codi feels unfit for any employment more meaningful than late-night Slurpee duty at a Tucson 7–11. When Doc Homer's deteriorating mental state leads Codi back to Grace to care for him and to teach high school biology, she feels terrified despite her extensive background in science and an all but completed medical degree.

Hallie, the other Noline sister, offers a stark contrast. Though the two sisters were inseparable as children and well into adulthood, they followed vastly different life trajectories. Codi, once the youthful firebrand, now drifts, bereft of ambition and passion. Hallie, who as a child had dutifully followed Codi on her missions to rescue coyotes and otherwise save the planet, has become the leader who dares to go wherever her beliefs take her. Hallie's itinerant idealism eventually leads her on a quixotic and ultimately fatal mission to aid Nicaraguan peasants under attack by American-funded Contras. The novel's treatment of Hallie's behavior is markedly ambivalent. Though brave, her actions seem neither productive nor wise. Rather than address the problem—American unwillingness to tolerate an alternative economic system to the south—Hallie treats the symptom, with predictable results. She dies a martyred hero, and the dirty war continues.

Codi, on the other hand, when confronted with the local mining company's poisoning of the land and water of Grace, does not follow her usual routine of abdicating responsibility and fleeing the area. Doc Homer's progressive degeneration holds her in place. Instead of leaving, she stays and attacks the problem from within. This reengagement with her surroundings and community enables a rediscovery of self. To her surprise, Codi finds her sense of self integrally tied to the health of the society and ecosystem. The danger to Grace resulting from the mining company's actions and the silent complicity of the EPA catalyzes Codi's long dormant value system, which cannot tolerate environmental degradation.

Upon returning to Grace, Codi wanders through the local orchards, mentally comparing them to graveyards and remarking on the excessive fall of fruit. Her observation is portentous because the orchards occupy an important niche in the town's cultural tradition, and their yield provides a livelihood for the town's inhabitants. Codi's biology class later discovers that the river running through Grace verges on biotic death, its pH level akin to battery acid. Years of copper mining by the Black Mountain Mining Company (until a few years before, the major employer of the town's men) created an enormous tailings pile[6] through which the company now leaches sulfuric acid in order to recover trace amounts of valuable metals. The acid seeps into the river, poisoning it and inflicting slow death on the crops of local farmers and orchard growers. The situation is both symbolically significant, as a metaphor for the hydraulic disenfranchisement of minority cultures (in this case, the indigenous and Hispanic communities), and imminently perilous because the crops that feed the community are dying.

When Codi naively assumes that a report to the local authorities will halt the leaching operation, Viola, the town matriarch, quickly disabuses her:

> "Dam up the river," Viola said. "That's all they have to do to meet with the EPA laws. Dam it up and send it out Tortoise Canyon instead of down through here. . . . [I]f Black Mountain dams up the river, it's out of the jurisdiction of the Environmental Protection Agency." (*Animal* 111)

This exchange underscores a dangerous characteristic of systems that contain multiple function systems: complexity to the point of obfuscation. Because the government requires popular support to maintain its power base, its laws must offer a veneer of equity. Yet the state also depends on capital inflow for its autopoiesis. The challenge therefore lies in creating laws that do not incite disfavor among the rank and file yet buttress the corporate capital mechanism.

With water forming the arid region's privileged sign, much systemic complexity arises from the state's double duty of appeasing the water-poor while servicing the water-rich. To accomplish both objectives, the state must disarm its regulatory authority. It does so by writing laws that protect its less powerful citizenry and then legislating the means to cripple those laws' enforcement. Therein lies an invidious logic: it is illegal to poison a town's water supply but legal to desiccate the area. Similar hydraulic subterfuges have enjoyed a long tenure in the West, dating back to the pre-Powell era.[7]

Grace's problems also stem from its residents' prior forfeiture of their water rights. Uneducated and ignorant of the complex field of water law, the town's inhabitants happily sold these newly created abstractions that had no place in their culture. They were comforted because the buyer was the Black Mountain Mining Company, a welcomed addition to the community as its new major employer. As a result, according to Viola, "Nobody around here's got water rights. All these families sold the water rights to the company in 1939, for twenty-five cents an acre. We all thought we were getting money for nothing. We had us a *fiesta*" (*Animal* 111).

Here we see clear echoes of the Owens Valley, where Fred Eaton posed as a prospective rancher and bought options on the best water rights in the valley. He then sold those rights to the city of Los Angeles at a huge profit. His actions are fictionalized but wholly recognizable in *The Ford.* The theme of purloined or misappropriated water rights is common in both historical and fictional representations of the West.[8]

In *Animal Dreams,* when the mining company diverted attention from the water to the abstract field of water rights, it shifted the discourse into a realm where it held the advantage. The company is an outgrowth of the society that invented water rights; the indigenous community struggles to survive on that society's periphery. Before the Anglos came to the region, water had been apportioned according to its use value and a historic sense of community. When the state assigned water an economic value and named it a unit of exchange, it affirmed water's place as privileged sign and solidified the economic system's control over it.[9] Once control over water passed to state and corporate interests, it cast grave doubt on the long-term survival of indigenous communities.

The EPA's impotence in the face of Grace's imminent demise has multiple historical referents as well. As a regulatory arm of an apparatus designed to facilitate capital gain, the EPA must constantly renegotiate its joint allegiance to corporate ideals and environmental protection. Since the social system is driven by self-interest (autopoiesis), it strives to defend those qualities that serve its aims. However, as the system diversifies, its aims become disparate, and conflicts arise between various function systems.

An excellent illustration of the regulatory tightrope the EPA must negotiate involves the ongoing controversy over pesticide levels in food. Environmental groups and concerned citizens advocate strict limits, arguing that the pesticides are toxic and that any exposure in humans constitutes a health hazard. Agricultural interests counter that trace amounts of the chemicals do not harm humans and are essential to the business of growing food. Faced with competing demands from constituencies that are both vital to the survival of the social system, the EPA must decide what constitutes a safe level of pesticide in food. That level must appease public health advocates without crippling the food production industry.

Even among potential allies, complexity makes conflict inescapable. A rafting concern and a hydroelectric company may wish to use the same river, but their respective commercial aims clash. If the river is dammed, it loses allure to potential rafting clients. Yet, without a dam, the power company cannot generate. Both businesses seek to exploit the resource potential of the river, but their methods are incompatible. Determining how best to manage the river becomes political, a process of reconciling the multiple domains comprising the system's self-interest.[10] As societies discover or define their environment's ecological limits, they are forced to take measures to guarantee their population's continued survival in the global political economy by becoming "environmental protection agencies" (Luke, "On Environmentality" 69). They accomplish this partially through determination of system boundaries.

Boundary drawing is an inherently subjective process, infused with ideology and integral to the distribution of power. System boundaries exert a strong influence on language, which, as noted earlier, is integral to the system's autopoiesis. One of the most potent weapons that language allots to the dominant class is the power to rationalize and justify its behavior. Ethics are, after all, a mutually agreed upon set of values determined in part by the configuration of the group setting the values. If the dominator can plausibly claim that its actions are morally just and cause no harm, then the complaints of the subjugated lose legitimacy. Their indignation excites no sympathy, and their complaints fade into insignificance. Power over language also permits ecological degradation because that degradation is a boon rather than a bane for those who benefit from it. In Grace, for example, the Black Mountain Mining Company's publicist might describe the acid leaching operation as a "recycling" of the tailings to extract surplus value from already processed material in order to provide the greatest possible return on shareholders' investments.

Boundary drawing can also affect one's race, even one's humanity. Over the course of recorded American history, from the Puritans through the present day, dominant cultures have subjugated other cultures or races for their own gain through redrawing the boundaries of the human to exclude the targeted group. Slavery is perhaps the most obvious instance, but other examples abound.[11] Before turning to specific events in Kingsolver's town of Grace, it is worth considering the basis for the power that inheres in boundary drawing.

System boundaries are cognitively set, and erecting them *eo ipso* sets the environment's borders as well.[12] The environment contains the innumerable other systems that comprise reality. Defining the environment also subjectifies it; my environment must and will vary from yours because we comprise two discrete function systems (even while forming parts of the larger social system), and therefore we form parts of each other's environment. If environments vary with subjectivity, so too must the linked concepts of pollution and environmental protection. Pollution "involves questions not only of concentrations but of consequences" (Evernden, *Social Creation* 4). That fact reveals moral and empirical components to the term, as well as to the problem it describes. Identifying pollutants involves determining that a foreign presence and a potential source of harm exists within the system.[13] Making this determination requires two potentially problematic steps: designating the system's boundaries and defining harm. As might seem natural given the nature of these issues, international law has often provided the medium for their exploration.

One of the most influential pollution-related disputes in international law is the Trail Smelter Arbitration (1935–41) between the United States and Canada. The arbitration arose from a cooperative effort by the two nations to mitigate the damage caused by airborne pollutants that had crossed into the United States from Canada. A principal problem facing both the parties and the arbitrators involved the lack of consensus definitions within the international community for pollutant and damage. Every country had its own definition, and the meanings were often at odds, rendering international regulatory agreements nearly impossible. Crippled by the lack of legal precedent or a governing body of law, the arbitrators cobbled together a decision that rejected virtually all of the U.S. claims.

The Trail Smelter was built in 1896 in Trail, British Columbia. During the ensuing years, emissions from the smelter drifted across the border into Washington and fell in the form of acid rain and acid fog, causing substantial property damage. In 1931, a joint U.S.-Canadian commission concluded that the smelter had caused $350,000 worth of damage in the United States. In 1935, following complaints from residents of Washington, the settlement was deemed unsatisfactory and resubmitted for arbitration.

In 1941, the arbitration tribunal ruled that foreign emissions did not constitute pollutants unless and until they caused actual, provable, and substantial damage. If an injury could not be measured in monetary terms, there was no damage and, hence, no remedy at law. Consequently, the United States received no compensation for violation of its sovereignty (trespass across its borders) by the smelter's noxious fumes because the loss was intangible. Similarly, the tribunal refused to hold Canada liable for damage to urban property in the United States because "there [was] no proof of facts sufficient to estimate the reduction in the value . . . of such property" (*Trail Smelter Arbitration,* United States v. Canada, 3 U.N.R.I.A.A., 1908).[14]

The Trail Smelter decision is still relied upon in international law. According to its holding, harm to a nation's air is not compensable under international law. Nor can a country seek damages when foreign emissions harm wildflowers, birds, or any other resource that, while priceless, has no assigned monetary value. The rationale for the decision stems from the tribunal's ill-conceived attempt to assign fixed definitions to value-based and mutable terms like *pollutant* and *damage.*[15]

Consider again the West's hydraulic dilemma: environmentalists decry profligate expenditures of water and the degradation of habitat caused by the damming of rivers, while supporters of agribusiness and continued western development defend the expenditures as necessary and the dams as beneficial. "To the environmentalists, what is at risk is the very possibility of leading a good life. To the industrialists, what is at risk is the very possibility of leading a good life. The debate, it appears, is actually about what *constitutes a good life*" (Evernden 5). The good life, in any system, involves keeping pollutants at minimal levels. Unfortunately, as in the Trail Smelter affair, systems differ on what constitutes a pollutant.[16]

In *Animal Dreams,* Grace's inhabitants view the mine's leaching operation as pernicious to the community and to the region. Residents who were once part of the Black Mountain Mining Company have become external and superfluous. The company's new configuration has forced a redrawing of systemic and environmental boundaries.

From the company's perspective, poisoning the river with acid amounts to an insignificant side effect of a beneficial process; the river's biotic death poses no danger to the mine's viability. The townspeople, meanwhile, derive their cultural identity from the river and, following their exclusion from the economic sustenance the mine had provided, rely on it for food and livelihood. The opposing views of the mine and the townsfolk reiterate the flexibility of the term *pollutant*. The mining company considers the sulfuric acid an asset (and the river extraneous). Grace's residents, by contrast, view the acid as a pollutant (and the river as essential). If pollution means "matter out of place," or a foreign object interfering with the efficiency of a given system, both sides are correct. Even when we attach a moral component and say that pollution is an "unnatural" contaminant that could cause terrible harm to a given ecosystem, both sides remain right.[17] Terms like *unnatural* and *harm* are creations of the systems that give them meaning. Their meaning derives from the myth system into which they are integrated.[18]

From Black Mountain's perspective, the acid serves a vital purpose: aiding the continued prosperity of the local function system (the company), the larger function system (the economy), and the overarching system (society). The town advocates a policy aimed at safeguarding the viability of the land and water, as well as respect for the diverse cultural heritages that make human society unique. The strife between Black Mountain and townspeople pivots on the differing perspectives from which they attempt to facilitate autopoiesis. Each side maintains that its vision best serves the long-term needs of the system. Since both predicate their arguments on sustainability, it becomes clear that sustainability is a contested term whose meaning does not always coincide with an "environmentalist" agenda.

> Sustainability . . . cuts both ways. On the one hand, it can articulate a rationale for preserving Nature's biotic diversity in order to maintain the sustainability of the biosphere. But on the other hand, it can also represent an effort to reinforce the prevailing order of capitalistic development by transforming sustainability into an economic project. . . . Sustainability . . . becomes a discourse about exerting power over life. (Luke, "Environmentality" 28)[19]

Each system, in defining its boundaries, tacitly articulates its autopoietic requirements. When the needs of two function systems conflict, as with Grace and the mining company, the system best able to withstand the perturbation caused by the conflict will gain influence and power in subsequent interactions. This concept of realignment through conflict comprises a crucial subtheme of *Animal Dreams* as well as a seminal component of systems theory. And since the novel is set in the West, the nexus of the conflict is water.

For Black Mountain Mining, the ability to poison Grace's river bestows tremendous power. Perhaps more importantly, the company's ability to waste water in the pursuit of trace amounts of metals indicates an altered, but still potent form of faux potlatch. After wasting the water and fouling an entire aquatic ecosystem, the company turns to the EPA for permission to dam the river and permanently remove the water from circulation. The dam will increase water's exchange value, swell the worth of the company's assets by facilitating further extraction of valuable metals from the tailings, and abrogate the community whose objections to the poisoned water caused a slight catch in the company's engine of commerce.

In essence, Black Mountain seeks government assistance in constructing a dam that will solidify its cycle of lucrative hydraulic profligacy.[20] Given the circumstances—a small, minority community wriggling in the jaws of a monolithic corporation—few would gamble on Grace's chances for survival. This is particularly true in light of the EPA's silent complicity in the company's scheme.

EPA's role in the dam's approval and the town's demise is a by-product of the double bind facing government agencies. The agency's primary duty, as a regulatory arm of a system dedicated to extractive commerce, is the facilitation of that commerce. To the extent that economic growth dovetails with the nation's ecological well-being, so much the better. But when those goals conflict, capital inflow remains the system's priority.[21] Nevertheless, the EPA, as its name suggests, is also responsible for stewarding the nation's environment. The conflict within its dual mission reflects a basic incompatibility between ecological sustainability and economic growth.

According to William Ruckelshaus, who headed the EPA during the

Nixon and Bush administrations, "the strongest supporters of a forceful EPA are the industries it regulates. They want the government to set reasonable standards and they want the public to know that they are being enforced" (qtd. in Merchant, *Radical* 161). The key word in Ruckelshaus's statement is "reasonable." Reasonable standards vary with the agenda of the observer. Rational regulations, in industry's view, soothe public unease without constraining corporate growth. Environmentalists, meanwhile, advocate standards that safeguard the ecosystem, with profit margins forming an ancillary concern.

While the goals of the two sides differ fundamentally, there does exist a middle ground where a temporary compatibility becomes possible. Those in industry want to seem ecofriendly so as not to alienate their client base and to postpone fouling their own nests. A reputation for "green" behavior often benefits a corporation's bottom line. Into this mix one must factor the desire of environmental groups to appear conscientious and principled, but also their need for significant capital infusions in order to function. These corporations and advocacy groups share a primarily white, middle-class constituency. Both industry and environmental groups can satisfy (or at least pacify) their support base through cooperation and public displays of environmental conscience.

Another factor spurring the uneasy cooperation between environmental groups and corporations is money. Corporate money in large part subsidizes the environmental movement. For example, the Big Ten environmental organizations[22] draw much of their funding from corporate donors. In some cases, corporate donations comprise more than half of their budgets. Many of the largest contributors are also infamous villains of the environmental movement, including Amoco, ARCO, Dow, Dupont, Exxon, Monsanto, and Weyerhauser (Tokar 16–17). The Big Ten's corporate donor base endows them with the resources to lobby effectively and reach a wide swath of the public. However, they also become beholden to an enemy agenda, dependent upon funds earned through the very activities they condemn. Their own autopoiesis depends on the system they criticize. It is a Faustian bargain to be sure, but a commerce-driven society offers few alternatives. That lack of electives epitomizes the central dilemma of Restoration as an alternative to Reclamation, as well as that of the environmental movement as a whole. Without a workable alternative to the current system, proponents of social change must ward off the most pernicious ecological assaults without sabotaging the economy.

It would seem that in a perfectly functioning commercial society, industry could police itself. Self-regulation of this type is impossible, however, because systems cannot predict the behavior of the environment. Systems adapt/react to environmental disturbances in order to reestablish equilibrium. Prediction, by contrast, requires insight into the structure and behavior of the environment and the ability to communicate that insight. Yet the environment, by definition, is that which the system does not know. Prediction therefore has no meaning in the context of system-environment interactions.[23] The reaction of a system to a given perturbation is a function of its structure and the nature of the perturbation. Furthermore, prediction does not exist outside of language, and language does not exist outside of the system. Since the cognitive act is linguistic and therefore communicative, it constitutes interaction within the system about the environment, not communication with the environment itself.[24]

Maturana and Varela draw an intriguing parallel between the phenomenon of trophallaxis (the conveyance of signals among insects through chemical exchange, as with ants coevolving to carry out different tasks within a colony) and "linguallaxis," the human ability to conserve adaptation and preserve social unity through language: "Because we have language, there is no limit to what we can describe, imagine, and relate. It thus permeates our ontogeny as individuals: from walking to attitudes to politics" (Maturana and Varela, *Tree* 211–12).[25] This dynamic occurs within social systems because their complexity enables self-observation.

The inherent reflexivity within the communicative act cannot avoid influencing the outcome of system-environment interactions. Because the system communicates about its response to a given perturbation, that communication will inevitably affect the system's behavior. Following Heisenberg's uncertainty principle, the position of the observer affects the outcome of the experiment. In this instance, though, the observer is also an actor; the actor is observing itself.[26]

Interaction between system and environment, by contrast, is necessarily nonlingual because the environment lies outside the system's boundaries and is therefore unknown.[27] Communicating with the environment via language (rather than simply reacting to disturbance) would require a common context from which to draw metaphor and thereby enable mutual comprehension. If a common context existed, that would involve shared experience, which would in turn require familiarity—and familiarity with the unknown is oxymoronic.

The dynamics of a system's relationship to its environment is nevertheless integral to its evolution. Every unity, whether a single entity or a society, looks to facilitate its autopoiesis. Every external perturbation, so long as it does not destroy the unity, causes it to respond and adapt. Without perturbation, the unity would stagnate and evolution would cease.

Accepting the relationship between environmental disturbance and evolution leads to two intriguing conclusions. First: natural selection does not accurately describe the evolutionary process because no selection actually occurs. Response is constrained by the type of perturbation and the structure of the system. Unities that do not respond in a manner consistent with continued autopoiesis do not survive, whereas those that successfully conserve their autopoiesis produce a next generation. For each system and each disturbance, however, there is only one possible response. All other options exist only in the mind of the observer. Second: the fact that external perturbation is beneficial calls into question humanity's historic tendency toward subjugating and controlling its surroundings. Ongoing attempts to tame "nature" and the "wilderness" indicate a disturbing devolutionary trend. These concepts will be explored in more detail in the concluding chapter. First, we turn to the question of how the dynamics of system-environment interaction manifest themselves in the fictional world of *Animal Dreams.*

The Stitch and Bitch Club

In Grace, the Black Mountain Mining Company maintains that opponents of the proposed dam must be neutralized because they create unwelcome resonance within the system (with resonance defined as the communication generated in response to a perturbation). Recall that environmental problems do not exist until they generate communication.

Environmental disturbance is relevant in direct proportion to the resonance it generates (Luhmann, *Communication* 29). Black Mountain can eliminate whatever impact and resonance Grace's citizenry might create by simply removing the disturbance. From its perspective, however, the disturbance is not the dam but the town.

Removing the town would serve the dual purpose of destroying the power base of the grassroots, land-based resistance and eliminating it as a subject of further communication. Without resonance, the river's contamination will cease, regardless of the pH of the water, because no one will complain about it. The inhabitants of Grace must therefore fight a two-fronted battle; they have to combat the river's contamination while simultaneously struggling to maintain their own societal legitimacy.

Codi's coming of age and subsequent participation in the saving of Grace stem from her liaison with Loyd Peregrina, the unknowing father of her miscarried child, and from her bonding with the women of Grace. The women attract her through their grassroots, activist ecofeminism, while Loyd offers a link to Codi's past via his spiritual, land-based values and tangible tie to her most traumatic period. Loyd's relationship with Codi benefits both of them. He learns that his values cannot be applied selectively and that all nature's creatures deserve to be treated with dignity.[28] She, meanwhile, learns of the spiritual power of the land and its ability to command an allegiance unlike that formed between sentient creatures.

> "Is there anything you know of that you'd die for?" I asked Loyd.
> He nodded without hesitation.
> "What?"
> He didn't answer right away. Then he said, "The land."
> "What land?"
> "Never mind. I can't explain it."
> "The reservation? Like defending your country?"
> "No." He sounded disgusted. "Not property. I didn't say property." (*Animal* 122)

Loyd explains by showing Codi places that "did not so much inspire religion as . . . seemed to be religion itself" (*Animal* 210). Loyd wishes to integrate reverence—an unquantifiable, yet undeniably present factor within the system—into people's everyday relationship with the land. He wants to integrate the land's sacredness into the economic and political systems, an act that would require renegotiating the boundaries of the sacred.

Viola and the other women of the "Stitch and Bitch Club" also demand a radical reordering of the system hierarchy but their methods are more immediate. Codi and Hallie's sardonic name for the coterie aptly characterizes the tension between the townswomen's traditional sex roles and their growing discomfort with them. It also describes their willingness to confront the language buttressing the status quo. The women, especially Viola, are acutely conscious that their movement's success depends on a radical reworking of their vocabulary:

> "Don't call that company the Mountain," she said curtly. "It makes it sound like something natural you can't ever move."
>
> "I've heard the men call it that," I said.
>
> Viola snorted like an old horse and started up the hill. (*Animal* 162)

Without constant vigilance, Viola knows that language can and will undermine the movement. Language is an extraordinarily powerful creative force. It builds societies and constructs ethical systems.[29] Ceding its power to an opposing social force can cause dangerous and irreparable consequences. Here, accepting the Company as a part of nature would capitulate to a vision of nature wherein acid in the water is not a pollutant and where local customs are disturbances to be quashed rather than a precious cultural heritage.

When the women meet to discuss options for saving the town, Viola makes clear that the success of the enterprise cannot depend on men (*Animal* 178). The initiative must come from women because women have shared the land's subjugation and therefore understand more intimately the necessity for action. Here, the rhetoric again recalls Mary Austin as well as the branch of ecofeminist thought that celebrates the woman-nature/man-culture juxtaposition as source of female empowerment.

One anonymous Stitch and Bitch member sums up the gulf between the male and female perspectives: "These men don't see how we got to do something *right now.* They think the trees can die and we can just go somewhere else, and as long as we fry up the bacon for them in the same old pan, they think it would be . . . *home*" (179). The trees' dying symbolizes their vanishing way of life, which, due to the narrowness of the dominant idiom, can only be described through nonlingual imagery. Nevertheless, the trees' role as metaphor endows them with significant power, particularly as the town struggles to rally support in and out of the community. The other potent symbol in the passage is the term "home," which communicates the anguish of displacement and the frustration of communicating the ineffable.

The Grace women's practical ecofeminism rejects the inculcated passivity that would permit the destruction of their homes and histories. Instead, they look to women-defined values that maintain that nonhuman nature, as the source for all life, must supplant the profit motive as society's defining characteristic. The women's activism merges with the spiritual reverence of the Native American tradition symbolized by Loyd, whose Apache heritage is both matrilineal and matrifocal. Activism and spirituality also merge in the body and mind of Codi, who uses her fluency in the rhetoric of science to further the cause. Working as a community, the women of Grace, including Codi, successfully stop the dam and the acid leaching that had all but destroyed the town.

The women save Grace through folk art. They fashion peacock piñatas and sell them to raise funds and call attention to the town's battle against the mining company. Their art creates significant resonance within the larger society, capturing the imagination of Sean Rideheart, a Tucson art dealer experienced in manipulating the system's regulatory apparatus.[30] Rideheart convinces the Stitch and Bitch club to agitate for Grace's placement on the National Register of Historic Places: "He knew all the ins and outs of becoming a historic place. He explained where to begin, and where to go after that, to see that the river would run clean and unobstructed. There was a fair amount of bureaucracy involved, but the process was reasonably speedy." Once listed as an historic site, the town need no longer fear "the onslaught of industry" nor "demolition or other negative impact" (*Animal* 277).

Invoking government regulations to protect the town offers an ironic contrast to the legislated impotence of the EPA that permitted the problem to escalate. When the town gains protection, its newly historic status is doubly ironic because it forces the system to protect the point source of its disturbance—like a lion donning a sock to protect the thorn in its paw—and because historic status implies an important role in the evolution of the social structure. Evolution, as we saw earlier, results from disturbance. Historical importance derives from a significant role in the society's evolution. It follows then, that an event's place in history is directly proportional to the severity of the disturbance it creates.

Consequently, Grace's historical importance results from its active subversion of society. This means that Grace's attempt to undermine the social system wins it official recognition as an important component of the continued evolution and preeminence of the status quo. Society retains its hegemony by co-opting Grace's victory as a means of entrenching its own power. In other words, the women win the local battle but the victory costs them the war.

The inadvertent lesson of the novel seems to be that active subversion and agitation for social change only strengthen the power structure. This seemingly unsolvable paradox subordinates Grace's successful ecofeminist/ecospiritual, grassroots revolt to Codi's reawakening, her union with Loyd, and their subsequent procreation. Kingsolver structures the novel's denouement to downplay the ecosystem and stress instead the successful reproduction of the community. In designing the novel this way, Kingsolver tacitly affirms Luhmann's hypothesis that ecological problems only become so when they generate systemwide resonance. The fight against Black Mountain Mining engendered enough disturbance to bestir the capitalist communicative apparatus and force acceptance of a less profitable alternative. But the revolt could not replace the dominant system with a new ecologically sensitive one. The struggle generated little resonance outside of Grace and therefore offers no broad alternative to the western ecological dilemma.

Grace's historic status offers no long-term implications for social reform of western land or water policy. It just spares one town a dismal fate. The social system's autopoiesis still requires short-term capital gain. Similar ecological degradation will inevitably arise elsewhere, with less palatable results. This prospect tempers the elation one might otherwise feel at Grace's salvation. Small victories in the face of larger disasters offer at best a pleasure tempered by a grim awareness of things to come.

This overarching sense of dread is an interesting and probably unanticipated by-product of manufacturing a happy ending to this environmental novel. Given the current constraints, a happy ending must involve either removing the environment from its central position within the plot, as Kingsolver does, or offering a coherent system and discourse to replace extractive capitalism. Muting the environment within the plot strips it of the ability to generate resonance and thereby force evolution. The alternative, keeping it central and articulating an ecofeminist steady-state option, remains a grail—outside the realm of effective communication.

Attempts to formulate a steady-state alternative have so far lacked specifics. For example, Herman Daly suggests that a steady-state economy is "an economy with constant stocks of people and artifacts maintained at some desired sufficient levels by low rates of maintenance 'throughput'" (qtd. in Merchant, *Radical* 37). A steady-state economy would use the lowest possible levels of materials and energy in the production phase and release the smallest possible amount of waste during the consumption phase. Terms like "desired" and "sufficient" invite a degree of subjectivity that leads to abuse and multiple definitions, much as in the earlier discussion of pollution. To further complicate matters, a steady-state economy needs a consensus definition of an optimal standard of living—a goal that smacks of utopianism.

It seems that Kingsolver had little choice but to divert readers' attention from Grace's relatively insignificant ecological victory to the personal triumph bound up in Codi's self-realization. Without a working definition of environment that does not conjure images of an extractive, ecologically pernicious economy, or an unquantifiable "not us," there can be no ecologically sanguine happy ending. Even the definition of ecology posits a barrier between organisms and their environment by semantically solidifying the schism between the two.[31] We define ecology as the science treating the interaction between organisms and their environment. Interaction contains an implicit alienation—it requires two discrete entities. The environment is therefore everything "not me."

A cybernetic approach, however, dissolves the barrier between self and other, even as systems theory dictates that the environment encompasses all that cannot be known or communicated. Mary Douglas points out that "it is not just the environment that is at risk, but the very *idea* of environment, the social ideal of proper order. The power which presents a danger for careless humans is very evidently a power inhering in the structure of ideas, a power by which the structure is expected to protect itself" (113). The destabilization of the term *environment* in turn undermines common notions of ecology. The resulting linguistic uncertainty could facilitate a new rhetoric of environmentalism more responsive to alternative ideological and economic systems.

Kingsolver's attempt in *Animal Dreams* to create an ecofeminist, organicist environmentalism does not succeed because, as yet, there exists no coherent vocabulary to describe the new regime. Without that vocabulary, combating initiatives like mines and dams must take the form of negative constructions couched in the language of the dominator. We can only say "No mine" or "No dam"; we cannot offer a coherent alternative regime. That strategy is necessarily reactive. It requires waiting for someone to suggest an act of environmental degradation before formulating a strategy designed to maintain the status quo. Yet the status quo is not satisfactory; it degrades the land and privileges extraction and profit over ecological and social health. Environmentalists thus find themselves in a vexing double bind—they must fight to safeguard a system and situation that they condemn or face an even bleaker alternative.

Environmentalism might then best be described as a passionate struggle to maintain an unacceptable state of affairs. Under the current system, change too often means serious harm. Rather than agitate for change, environmentalists must impede it in order to delay the inevitable slide into ecosystemic collapse. This mission, simultaneously both hopeless and crucial, infuses the environmental movement with a combination of urgency and despair.

Such a grim and unrewarding struggle does not easily inspire optimism. Kingsolver nevertheless wished to write an environmental novel with a happy ending. To do so she either had to create the vocabulary to describe a sustainable reality or divert readers' attention to interpersonal developments. Of necessity, she chose the latter.

6

Imagining Sustainability

This is not the Age of Enlightenment, but the Age of Not Knowing What to Do.
—Walker Percy

Restoration's undefinability renders it unattainable even as Reclamation continues its assault on the nation's hydrology and geography. It seems there is no sustainable alternative. Happily, that is not the case. The path to sustainability may be hidden and overgrown from years of disuse, and probably does not lie where the map says it should. Nevertheless, it is extant and navigable and will take us where we need to go—if we can find it.

The path does not lie where it once did because the terrain has shifted. The Reclamation era has irretrievably altered the western landscape, causing what Neil Postman calls an "ecological change." An ecological change occurs when one segment of the whole metamorphoses so dramatically that the entire society changes (18). Postman cites the printing press and television as two inventions that wrought ecological changes. On a smaller scale, the dams, channels, and reservoirs of the Reclamation era have generated a similarly irrevocable shift in both the physical and the metaphorical landscape of American society.

While driving through California's Mojave Desert recently, I passed a sign advertising "Rain for Rent." This rain could be had at the Irrigation Store, an establishment dedicated to providing farmers with the tools for trouble-free reclamation at affordable prices. Taken in the context of a century of unprecedented technological advance and national obsession with the myth of the garden, an irrigation store does not seem particularly remarkable. In the larger context of human history, however, its existence is nothing short of staggering. Empires and countries, from ancient Chinese dynasties through the current nation-states of the Middle East, have risen and fallen because irrigation technology spawned hydraulic empires in which the ability to deliver water rested in the hands of the educated, wealthy few. Today, anyone can purchase the tools of dominion at a roadside store outside of Bakersfield.

The Irrigation Store was conceived out of hubris and enabled by the ecological change wrought by the Reclamation era. The store's existence testifies to the fact that the nation's geography, for better or for worse, is forever different. Restoration's goal of remaking pristine ecosystems therefore looms as an impossible goal, comparable to unmaking glass into sand. Furthermore, if healthy ecosystems exist in a constant state of flux, which "pristine" state should they reoccupy? Inevitably, such decisions become more ideological than ecological.

In a sense, the Reclamation era represents the turbulent adolescent years dividing this nation's infancy and its maturity. In the years following its inception, the United States expanded westward, and its resources and possibilities appeared limitless. Myths like those of the garden and frontier assumed that there *should* be infinite abundance but did not investigate whether there actually *were* such riches. Visions of abundance persevered despite mounting ecological evidence to the contrary. When the frontier closed and limits to growth became impossible to ignore, the Reclamation era was born. It embodied both a reaction to and an elision of limits to growth. Reclamation became a juggernaut, wreaking havoc on the western landscape in an attempt to further a bankrupt ideology of infinite possibility.

The Reclamation era divides into five stages, with this book devoting a chapter to each. The first stage begins with John Wesley Powell's vision of a "reclaimed" West populated by family farms whose size and crops were determined by the realities of arid-lands agriculture. The next four stages are chronicled with the aid of novels, each of which reflects the social, historical, and hydrological contexts of their respective periods.

The Ford shows the heady optimism of Reclamation's early years and the clash between urban and rural constituencies even while both claimed to represent the same values. *The Grapes of Wrath* takes place during the heyday of the Reclamation era, when water, dams, and delivery mechanisms trumped all other concerns, including human rights and the health of the ecosystem. Corporate agriculture co-opted yeoman symbolism and used it to enormous personal gain. Conflict was unavoidable, as *The Monkey Wrench Gang* makes plain. Abbey's novel reflects a growing disillusionment with reclamation for its own sake and a willingness to privilege environmental concerns over economic benefit. Last, *Animal Dreams* dramatizes Reclamation's end phase. It depicts a population struggling to disrupt and replace the cycle of corporate gain, environmental degradation, and hydraulic dependency created by nine decades of Reclamation rhetoric. That brings us to the present. We must figure out what to do now.

All societies rely on common myths; we must assume that we share assumptions about the world in order for language and its embedded metaphors to function. These shared myths enable human interaction. Language is built on metaphor, and metaphor requires an assumed commonality of experience. Ludwig von Bertalanffy suggests that such "moral concepts as Freedom, God, Immortality, and Human Dignity are fictions but nonetheless of immense importance: for we have to behave 'as if' they were reality. . . . [T]he myths of tradition are fictions based on the mythical experiences of man and later invested in historical narratives" (67). Accepting this premise can prove terrifying. It suggests that the fabric of society is not made up of universally held human values, but merely illusions shared for the sake of convenience. When and if those illusions become inconvenient, and we stop behaving *as if* we share a common reality, then the societal fabric will unravel. Without a shared set of experiences and assumptions, language cannot function and communication becomes impossible.[1] And without communication, society will dissolve into chaos.

Allegiance to the notion of a shared, objective reality also has serious, negative ramifications. Attaching objectivity and incontrovertibility to a given web of myths eliminates the need to interrogate social norms. Assuming that there is one correct way of seeing the world shackles us to a particular interpretation of our environment. Freed from the need to question our behavior, we simply act, secure in the knowledge that we do the "right" thing. However, that interpretation which we accept *as if* it were true may, like Reclamation, have been designed not to advance the common good, but rather to enrich a select group.[2] In Neil Evernden's view, this ruse of objectivity is all but inevitable, as are its consequences: "[T]he tendency to practice the subterfuge of mythmaking is very understandable. In practical terms, it may very well afford us some measure of comfort by legitimating a belief in the certainty of at least a few features of existence and a few behavioral norms. But in the long run, it solves nothing, and has the added effect of drastically transforming . . . nature" (29–30).

Shared myths inevitably mutate as societies change. Returning to the examples of God, immortality, freedom, and human dignity, we see that the concepts of God and immortality have undergone radical revisions in the last 150 years, while the ongoing tragedies in Burma, Sudan, Tibet, and elsewhere demonstrate that freedom and human dignity are not universally valued. In the United States, many of our common myths revolve around the frontier, the garden, and the absence of a precolonial history. We cling tenaciously to this mixed and shifting bag of fictions that underlie the American Dream.

It is time to reconfigure those norms that we all behave *as if* were true. Societies, like ecosystems, must evolve or die. We must, in Bruno Latour's words, amend the key articles of our "discursive constitution."[3] I propose a rhetoric of Sustainability that draws on many of the mythic properties of Restoration and Reclamation while adding elements that they fail to capture.

Sustainability is less a blueprint than an organizing metaphor for environmentalism in the post-Reclamation, information age. It would operate as an autopoietic system, evolving and gaining complexity in order to ensure its survival. Using a hypertext model, Sustainability involves a radical shift away from binarism and exclusive allegiance to one paradigm. Rather than offering an ideological ur-text, Sustainability envisions a system of "linked causes" that de-emphasizes particular ways of seeing while allowing access for all. Whereas Reclamation focuses on humans (humans re-creating Eden) and Restoration concentrates on the natural (remaking the ecosystem into a pristine state), Sustainability allows conflicting views to coexist and, by deprivileging any particular one, permits all to contribute to the health of the ecosystem.

This nonbinary flexibility is crucial to surviving in a world interconnected by increasingly powerful modes of communication. One of Sustainability's strengths lies in its basis in existing social trends. Rather than proposing to overhaul human interaction, its logic is based on the concepts and rhetoric of the information age. The multiple authors and unstable narratives that characterize hypertext and information theory have already had significant impact on the world of literary criticism.[4] As the Internet enters more homes and information age concepts continue to infuse pop culture, the notion of society as a global web will become ever more bound up in cultural mythology. Even in the short time since it became popular, the Internet has demonstrated enormous potency as a cultural sign. The term *information superhighway* is already a cliché. Meanwhile, cyberpunk dominates science fiction, inspiring movies like *The Terminator, Lawnmower Man,* and *The Matrix;* presidential candidates have home pages; and MSNBC and other interactive media look to wed television and the Internet and thereby confirm hypertext's place in the cultural mainstream.

Sustainability draws on the tools and concepts of the information age, while its roots lie in the systems-theoretical concept of metabalance. Metabalance posits that a system must be fundamentally out of balance to achieve stability. If a system expends energy, it must replenish itself from external sources. Just as we eat to renew our bodies' energy reserves, so do social systems require sustenance from without for their continued survival. That means that systems are always out of balance. An in-balance system that did not expend energy would be inert and lifeless. Like any vibrant, healthy system, a sustainable society exists in a state of beneficial flux with constant inflow and outflow. It relies on disturbance and shifting perspective for its vitality and survival.

In the West, the social system feeds on food and energy produced through Reclamation. But Reclamation requires unlimited water, an ecological impossibility. As water reserves exhaust, society teeters on the verge of disaster. By contrast, a sustainable society would not become overreliant on one source of energy. Overreliance comes from privileging one particular method or way of seeing. Sustainability requires multiple perspectives, and functions through accommodation; no one constituency can subordinate opposing views indefinitely.[5]

Hypertext offers a useful analogy for visualizing the Sustainability, or linked-cause, approach to environmentalism. In a given hypertext, multiple hot links connect to other hypertexts, which then have hot links to other texts, ad infinitum. The amount of information and the possible ways of ordering it are infinite. Participants create the text out of a maelstrom of unfiltered information and give it meaning by drawing boundaries and forging links.[6] The workings of the environment can be conceptualized the same way. For example, global warming links to fossil fuel combustion, agriculture, and deforestation. Agriculture, deforestation, and industrial pollution all link to each other, and each also links to myriad other issues. Each cause has its own constituencies with competing methodologies for restoring equilibrium. Nevertheless, each cause cannot be severed from the larger matrix containing all other causes. Attempting to solve any one cause in isolation exacerbates other problems. If, for example, we isolate global warming and, for the sake of argument, limit its causes to the four mentioned above—industrial pollution, agriculture, fossil fuel combustion, and deforestation (see fig. 1)—any attempt to solve global warming without also addressing its linked causes would prove disastrous.

Let us assume that carbon dioxide is the principal greenhouse gas contributing to global warming. A large portion of the carbon dioxide released in the United States is a by-product of burning fossil fuels, which create energy that powers everything from paper mills to recycling plants. If the federal government were to attack the problem by mandating a 50 percent slash in all greenhouse emissions, the price of goods would go up, industrial production would plummet, and many businesses would close, putting millions of people out of work.

Setting aside the social and political ramifications of a large, disaffected electorate, the resulting strain on agriculture and the nation's forests would be enormous. There would likely be an urban exodus as millions of people suddenly required land on which to grow their own food. The East, with its denser population, would run into immediate problems of space. Eastern forests, only recently recovered from centuries of indiscriminate clearing and urban expansion, would again face destruction. The forest clearing would generate a feedback loop similar to the one taking place in the Amazon. The mass clearing of forests would mean fewer trees absorbing carbon dioxide even as their burning

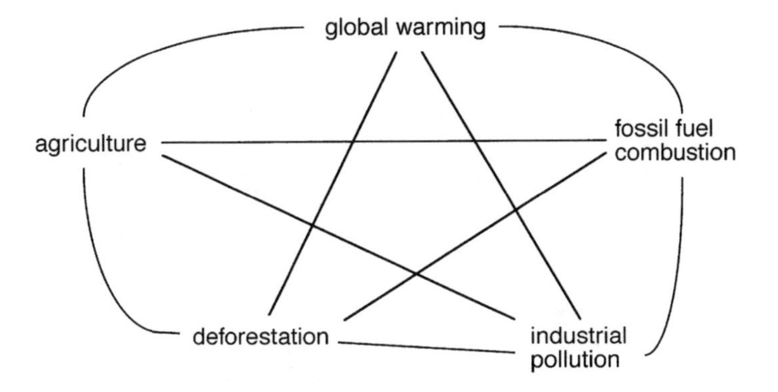

FIG. 1. An artificially closed system wherein global warming has four discrete causes, all of which link to each other and to nothing else

released more carbon dioxide into the atmosphere. Those conditions would accelerate rather than mitigate global warming.

In the West, space would pose less of a problem than water. We have already seen what resulted when legions of would-be yeomen flocked to the West and attempted to farm it. Now, with aquifers mined nearly to exhaustion and rivers oversubscribed, there would be even less water to go around. Furthermore, the sudden bulge in farmers, accompanied by a plunging GDP, would likely doom the country's existing agricultural base. Consecutive collapses of the nation's industrial and agricultural infrastructures would probably destroy the economy. Given the international community's interdependence with, and reliance on, the American economy, the resulting chaos could plunge the world into a protracted and bloody period of anarchy.

The above scenario is deliberately simplistic. A comprehensive analysis would have had to address the virtually unlimited possibilities and implications of one simple act (mandating a 50 percent cut in emissions) aimed at accomplishing one simple thing: capping greenhouse emissions in the United States. My point is this: no one cause or issue is separable from the larger web of causes to which it links. Any action at any given "website" ripples throughout the social system. Within the larger social web, however, as separate constituencies agitate for particular causes, the resulting microadjustments maintain a systemwide equilibrium. The key lies in providing access to the web.

Many of the Native American cultures that populated this continent before European settlers arrived and labeled it "virgin land" possessed a sane, forward-looking ecological perspective. Several, including the Mohawk and Iroquois, explicitly direct decisions to be made only after considering their impact on the next seven generations. In germinating our own sustainable myths, we would do well to consult and learn from those who have done it before. I am not recommending that the United States adopt Native American mythology wholesale, even if it were somehow possible to condense the hundreds of diverse Native American cultures into a how-to manual. No culture has a preexisting formula for healing the wounds of Reclamation. The solutions, like the problem, must be new.

The first step involves retooling the national mythology to incorporate the (relatively) recently acquired power to drastically alter our physical surroundings. This will entail integrating ancient perspectives of land use and history, acknowledging the much-multiplied impact of an increasingly technological and proliferating population, and developing an expanded systems-theoretical model that incorporates predictiveness as an autopoietic mechanism.

In 1872, John Wesley Powell declared that the actions of humans on the face of the earth were "trivial," and that "the conditions which they produce are of minute effect, and in presence of the grand effects of nature escape discernment" (*Report* 91). No more. In the interim 120 years, we have quintupled the world's human population (in the United States, it has sextupled), circumnavigated the planet at speeds exceeding that of sound, orbited, mapped, and so refashioned earth that we now talk in terms of "global" warming and the "end of nature."[7]

In the Middle Ages, behavior was strictly circumscribed through imprecation and allusions to the wrath of the Almighty. In the postindustrial age, we have not only lost our religious awe, we also seem unconcerned about the harm our behavior can inflict. When cruising the information superhighway, we would do well to peer out at the ravaged landscape and abandoned heavy equipment rotting on its shoulders. It has never been more important to respect and even fear our own potential and past.

We live, systems theorists believe, in a world of our own design, interacting with an environment that we can never understand. The multiplicity of perspectives that form our collective reality deters the privileging of any specific system or environment. At first, this theoretical framework appears to negate contemporary rhetoric proclaiming humans a part of the environment. But further consideration shows that systems theory does offer the possibility of commingling humans and the environment.

Systems are in constant interaction with their environments, which are themselves other systems. All exist in a continual cycle of perturbation and adjustment. Once a system learns to respond beneficially to a disturbance, that disturbance becomes integrated into the system. The system grasps how to communicate with and create meaning from the disturbance, thereby transforming it into a known quantity and removing it from the shadowy realm of environment. If you picture it in terms of a map, imagine the environment ceding a little territory to the system as a result of a border skirmish.[8] This boundary shifting occurs constantly and has important implications. Consider the following:

1. System boundaries exist only in the mind of the observer. Where, for example, does the tree root end and the soil begin? When does water cease being water and become blood? Where do I end and the bacteria in my digestive tract—organisms fully capable of existing in other media—begin? The quality of these questions suggests that no boundary between self and other or between system and environment can withstand prolonged, multivalent scrutiny.[9]
2. We are each other's environment. Even as humans are components of the larger social system, we also form autopoietic unities with discrete environments. Other unities populate those environments. One unity may form a part of many discrete systems and even more environments. The overlap between systems and environments, as well as their observer-determined status, makes

differentiating between humans and environment impossible and the attempt counterproductive. Once the artificial rigidity of these boundaries gives way to a negotiable border between environment and humanity, the environment becomes part of us and ecological conscience becomes an autopoietic imperative.

3. We cannot evolve without an ongoing relationship with the environment, including all the perturbations such a relationship entails. Consequently, the ideal of controlling nature, aptly encapsulated in the 1965 Bureau of Reclamation proclamation, "Man serves God. But Nature serves Man," is not only quixotic, but pernicious. Without unpredictable, uncontrollable perturbation, the human species would languish and die. In other words, Thoreau's overused but little understood saying, "In Wildness is the preservation of the World," continues to inform the environmental debate. We need wildness not to assuage an antiquated wilderness aesthetic, but to preserve our ability to evolve.[10]

The presence of permeable boundaries between system and environment and the need for an ecological conscience reintroduce prediction as a component of system-environment interaction. N. Katherine Hayles suggests that reality arises from an "unmediated flux" that gains pattern and form through interaction. It comes into being through mediating the flux via "self-organizing, transformative processes that include sensory, contextual, and cognitive components." Both the flux and our organizing powers are necessary to the construction of reality. The flux exists irrespective of us, but we give it coherence. Hayles labels the interactive process that shapes reality "the cusp" ("Ground" 49). "Riding the cusp" involves a healthy awareness that both the flux and constructed reality are central players in any interaction. In a reality constructed entirely through human thought, one could walk off a cliff and not fall, whereas in an objectively determined world, frogs and humans would perceive the same reality despite species-specific vantage points. Of course, neither scenario occurs in the world as we know it. The middle ground lies in what Hayles calls "constrained constructivism" (53), which acknowledges that if you jump off a cliff no amount of deconstructive posturing will prevent your fall, but also asserts that any number of culturally based explanations (gravity, God's will, etc.) could explain and give meaning to your descent.

Assigning meaning to an object plays a crucial role in that object's creation. The same principle holds true for the larger concept of environment. Since environment forms out of the flux, not wholly of it, interaction plays a crucial role in its construction. That interaction renders it partially known and therefore partially predictable as well. Hayles contrasts this vision with "Maturana's world," where the actions of autopoietic systems are determined solely through their reactions to disturbance and have no cognitive element at all ("Nature" 11).

Accepting Hayles's formulation of the subject/object, system/environment relation means that cognition plays a key role in each component's formulation. If human systems can predict or react knowledgeably to perturbation, the outlook for social change brightens. Predictive ability, combined with the enormous flexibility inherent in the multisystem approach, inspires a degree of hope for the hydraulically beleaguered southwestern United States.

Prediction requires a familiarity with history. Consequently, long-ignored Native American precepts, along with an awareness of the endemic national indifference to the repercussions of our myths and technology, can play crucial roles in devising a rhetoric of sustainability. We can and should consult our history of ill-gotten subsidy, faux potlatch, and desiccated aquifers when realigning and reaiming national environmental priorities.

Ever since the successful Sagebrush Rebellion against the ecologically sane but politically misguided policies of the Carter administration, western agricultural and political interests have ferociously resisted efforts to bring mining, grazing, and agricultural policies into step with ecological realities. During the early days of his administration, President Carter tried to institute fiscal and ecological restraints on numerous water projects. His proposals generated a tsunami of criticism in Congress and across the nation. Public Works Committee chairman Ray Roberts called Carter a captive of "environmental extremists and budget hackers" (qtd.

in Reisner 330). Nowhere was the storm of criticism more fierce than in the West, where politicians and citizens alike accused the Carter administration of attempting to destroy the western way of life.

In recent years, the rhetoric has taken a new turn that hearkens back to the halcyon days of the Reclamation era when appeals to the yeoman ideal invoked automatic federal subsidies. In 1994, Helen Chenoweth was elected to Congress, representing the state of Idaho. Congressman[11] Chenoweth is a fierce advocate of decreased federal presence in the West. She is also a founding member of the "wise use" movement, which advocates abrogating environmental laws and channeling increased power to property owners. One of Chenoweth's first acts following her election was to draft a bill she called the Civil Rights Act of 1995. The bill would have barred all federal agencies from enforcing any laws without permission of state and local officials.

Chenoweth's bill propounded a states' rights agenda that has led to armed confrontations between local militants and federal officials. The bill is neither an isolated example nor the most extreme articulation of the states' rights program. In a three-pronged political scheme whose internal contradictions are almost breathtaking, States' rights proponents advocate "returning" federal lands to the states (land that the states never owned; it was ceded to the federal government as a condition of statehood), reducing federal jurisdiction at all levels; *and* maintaining federal subsidies for mining, ranching, and irrigation so that the "western way of life" will not perish. "Get out and give us more money!" has never more accurately described this powerful component of western political consciousness.

States' rights advocates have also learned the lessons of the Reclamation era and seek to build on its legacy. Their rhetoric proclaiming the need to "reclaim" federal lands echoes the more explicitly water-based Reclamation myths that propelled the West into hydrological insolvency. Where Reclamationists looked to reclaim an imaginary garden from the desert, their ideological progeny demand the return of land they never owned from a federal government that they accuse of undermining the "American" way of life. As Chuck Cushman, a leading spokesman for the wise-use movement declared following the Republican electoral gains in

1994, "Bruce Babbitt tried to say there was a new West. The Old West voted in November [1994]" (qtd. in Blumenthal 30).

The fact that the Old West Cushman refers to never existed anywhere but in the American imagination in no way mitigates its power. Suburbia, to borrow Marshall McLuhan's phrase, "lives imaginatively in *Bonanza*-land" causing an influential segment of American culture to "march backwards into the future." Technological advances and the nation's gigantic groundwater reserves permitted Americans to spend the last century behaving *as if* the myths of the garden and the Old West conformed to the ecology of the arid lands. The rhetoric of Sustainability, by contrast, implies the possibility of a radically reorganized society even though the prospect of such a reorganization looms as all but impossible.

Still, I maintain that the task can be done. The key lies in successfully "riding the cusp" governing the interaction and construction of self and environment. This will require enhancing the system's predictive capability to offset the uncertainty that will accompany the demise of the old myths. That the human social system does not blindly respond to external perturbation, but rather constructs responses based on history and knowledge, means that autopoiesis is partly a conscious procedure and therefore subject to renovation. Language can enable that renovation process, placing social change within our reach.

History shows that we have made the reach in the past and that ecological changes have resulted. Postman observes that when evolution displaced Genesis as the myth of creation, it left a yawning epistemological void: "By calling into question the truth of [Genesis], science undermined the whole edifice of belief in sacred stories and ultimately swept away with it the source to which most humans had looked for moral authority" (160). That devastating outcome was attained almost exclusively through language. The mechanistic paradigm (which used the machine as the unifying metaphor for explaining nature)[12] through which seventeenth-century thinkers revised the myth of the garden is not yet four hundred years old, and Reclamation is only ninety-five. The biblical version of creation lasted thousands of years and until recently enjoyed the support of very powerful language and social systems. It nevertheless was successfully challenged by a language-based alternative that could not even boast the imprimatur of any central authority figure. In just a few centuries, human society unraveled a five millennia-old epistemological quilt. That is not to suggest that the Genesis myth does not retain considerable power and credibility. However, when compared to its power and authority in centuries past, Genesis has clearly lost a lot of ground to alternative explanations of creation.

Embedded within language lies unmatched power and potential to generate social change. The problem is that this potential lies fallow, allowing language's might to flow into system maintenance. Instead of trying to restore ecosystems to an imagined prior state, we might better devote our efforts to restoring language to its role as an agent of change. Reenabling language is a central aim of Sustainability, and, given the success of the mechanistic paradigm in supplanting Genesis, it is an achievable goal for the forces of environmentalism. Nevertheless, even within the systems-theoretical definition, environmentalism continues to lack focus because of the environment's inherently subjective qualities. The term's parameters have narrowed, however, and once we accept the premise of a shared system of overlapping environments, it becomes possible to hear voices of change, even when, as in *Animal Dreams*, the voices come from the traditionally disenfranchised.

Sustainability does not require that society metamorphose into an egalitarian Kismet. Complex systems require a complex web of relations. As with hypertext, not every local component or link interacts with every other one. The map of interactions determines the global structure, which in turn determines the status of the local components. As in any observer-defined ordering system, some hierarchy is unavoidable.[13] Nevertheless, since each interaction within the system changes the global structure, all components have access to the mechanisms of change. To use another computer example: over the last twenty years, a number of myths have developed concerning how the world of computers was entirely refashioned by hackers working out of their garages. One of those early hackers was Bill Gates. Now the blue suits at IBM speak hacker lingo because it maps better onto an information age society.[14] Hacker influence migrated vertically; voices from the periphery generated enormous resonance and forced the insular world of computing to respond and adapt. Though a clear hierarchy existed and still exists, bottom-up integration proved a viable mode of access.

One of the most compelling advantages of the systems-theoretical approach lies in this redefined notion of access. With fluid boundaries, no one way of seeing harbors any more inherent legitimacy than any other.[15] Jean-François Lyotard argues that the systemic trend toward complexity has no allegiance to anything other than complexity itself: it "is not attached to an Idea. . . . It assimilates risks, memorizes their informational value and uses this as a new mediation necessary to its functioning" (7). We are simultaneously components of a large number of systems and environments, human and nonhuman, all similarly situated. Since we are each a network of systems that all form parts of other systems, it is impossible to define the boundaries of self with any certainty or permanence.

In hypertextual terms, each function system forms a hot link that takes the observer from context to context. The sequence of links and the narrative they form are entirely observer defined. As the observer's perspective changes, the system and its boundaries must also shift. What was once environment becomes system and vice versa. In light of this transitive relationship between self and other, degrading the environment is tantamount to self-destruction, while protecting it becomes an autopoietic imperative. Viewing environmental protection as self-preservation conveniently removes it from the moral realm, eliminating the need for appeals to conscience to prevent its degradation. It also does not extend human status to other life-forms—a criticism often leveled at some forms of radical environmental rhetoric.[16]

When a cause is co-opted by the status quo, as in *Animal Dreams,* it strengthens the dominance of the current paradigm. If, on the other hand, thousands of disparate causes militate for specific changes that do not necessarily dovetail with a national agenda, the system cannot co-opt them all because they often conflict, not just with the system but with each other. Paradoxically, the greater and more varied the perturbations a system experiences, the greater the likelihood of the seemingly opposing consequences of drastic upheaval and long-term health. Only through constant, unremitting flux do systems achieve metabalance.

Analyzed in these terms, and from the standpoint of both elementary logic and ecosystemic reality, Reclamation fails. First, Reclamation requires the state to provide an unlimited supply of water. Water powers the system, keeping it in metabalance. But Reclamation's existence is predicated on correcting a scarcity of water. Clearly, water cannot be both scarce and infinite. This contradiction at the base of Reclamation's existence means that the primary values or beliefs underlying the state's authority in this area are fundamentally inconsistent. Recognizing this inconsistency causes a loss of faith in the central authority, leading to what Habermas terms a "legitimation crisis." Legitimation crises are particularly problematic for modern societies because of their reliance on reason rather than religion and tradition.

Second, the ecological ramifications of this incoherence include an inevitable and worsening water shortage. This is the specter now facing the West. In addition to the Colorado River's oversubscription, the Ogallala aquifer, which underlies Texas, Kansas, Colorado, Oklahoma, New Mexico, and Nebraska, is fast depleting. Through the invention of the centrifugal pump, the Ogallala's enormous groundwater reserves pulled the Midwest out of the Dust Bowl and reestablished the region as the world's breadbasket. Unfortunately, over the last fifty years, more water has been pumped out of the Ogallala than it managed to accumulate in the last five hundred thousand. A Lake Huron's worth of nonrenewable water has vanished, never to return (Reisner 11). The shortage is both real and imminent. Reclamation fails on its own terms because there can never be enough water to fuel its indefinite autopoiesis. If we continue to ignore this ecological certainty, the system will no longer experience the beneficial flux that characterizes metabalance. It will collapse.

Luckily, Reclamation is not the linchpin of the entire social system. Though Reclamation has rhetorically wrapped itself in a cloak of indispensability over the last century, it remains but one component of a much greater whole. Entrenching itself in western iconography allowed Reclamation to withstand the many voices—from Mary Austin to Barbara Kingsolver—pressing for reform. Yet those voices, long exiled to the periphery, are vital to maintaining metabalance. Deprived of their input, the system has grown dangerously insulated and slow to change. Language has also stagnated, rendering advocates of change increasingly inarticulate. In addition, the linked issues of an incoherent hydroeconomy and a nonresponsive sociopolitical apparatus have left the nation highly vulnerable to an environmental energy crisis.

The alternative to social collapse is evolution. Systems evolve by responding to the flux born of dissonance. If dissonance is stifled, the system stagnates. Progress therefore can come from amplifying the voices of groups presently disenfranchised. Before progress can occur, though, the concept of progress must be interrogated and revised. Currently, "economic progress" links to the notion of a "high standard of living." But, as Timothy Luke argues, the crises of everyday material life over the past quarter century have unmasked this myth and shown that the perceived high standard of living is not standard, rarely high, and may not even be living (*Ecocritique* 201).

Even without a finely honed lexicon or a truly proactive ecological strategy, Sustainability's precepts can beneficially influence public policy. With the Ogallala aquifer running dry, Bureau of Land Management lands turning into parched and polluted gullies, rice sprouting under six inches of standing water in the California desert, and the Colorado basin turning saline and desiccated, pushing for an end to hydrological and other environmental inanities serves an obvious and valuable purpose.

I write these words in Los Angeles, a city in perpetual hydrocrisis whose water is almost entirely imported from other arid regions. Yet I cannot leave my building without splashing through puddles created by ubiquitous sprinklers. Pointing out the problems and peculiarities caused by disparate ecological visions has been the task of the environmental movement virtually since its birth. Landmark victories like the Clean Water Act and the Clean Air Act, the Wilderness Act, and the creation and development of the national park network testify to the movement's ability to achieve notable successes even with a reactive vocabulary. However, as the movement has grown, its rhetoric has become diluted by the conflict between maintaining the health of the environment and preserving the environmental movement. The movement continues to stave off much industry and government-sponsored harm, but its reliance on corporate funding also raises issues of credibility and diplomacy.

In recent times, the NIMBY (Not In My Back Yard) phenomenon has endured scathing criticism as a symbol of selfishness and snide unconcern with larger social issues. While much of this criticism has merit, the NIMBY phenomenon still bears scrutiny. When residents of a wealthy, gated community object to nearby low-income housing, they might deservedly stand accused of isolationism and disregarding the needs of the less well off. But when members of a Latino community objects to a toxic waste incinerator going up in their midst, are they being similarly selfish? Inversely, if the Latino neighborhood agitates for the toxic waste incinerator to be built in its midst, because the incinerator will bring needed jobs to their region, what position should the national environmental movement take on the issue? The answer, I believe, should be no position at all. In the language of Sustainability, the linked causes comprising the environmental movement do not coalesce on this issue.[17] The presence or absence of the incinerator is a local concern, requiring local expertise and passion to solve.

Even within the environmental justice movement, itself a subgroup of environmentalism, different constituencies advocate different results. In this instance, the local community may believe that the economic boon the incinerator would bring outweighs its risks. That decision does not invalidate the environmental justice movement's larger goal of shielding low-income and minority communities from environmental hazards. It merely allows the diverse voices within the movement to be heard while acknowledging that local communities have important links to other causes (education, housing, job security, etc.), and sometimes those causes take precedence.

When a gated community rallies against a housing project, its members argue from a position of wealth and systemic dominance. Their political influence guarantees them attention at the policymaking level. Therein lies the crucial distinction between them and the minority community. As Kingsolver depicted in *Animal Dreams,* for marginalized groups, systemic resonance is hard won. In the unlikely event that underrepresented people generate sufficient resonance to force systemic adaptation, it rarely creates change beyond the procedural level. In *Animal Dreams,* the town of Grace stopped the mine from further poisoning the river, but the price of victory included strengthening Reclamation's grip on the region. That victory can be considered progress, but only in the sense that battlefield triage is good medicine: both require aggressive follow-up care to enable long-term survival.

Let me here draw an important distinction between the local empowerment I advocate and the misguided rhetoric of the states' rights and wise-use movements. Sustainability through local empowerment need not mean the demise or crippling of the central authority. That central authority (in this case, the federal government) can and should be integral in the maintenance of the region's environmental integrity. John Rawls notes that "unless a definite agent is given responsibility for maintaining an asset and bears the loss for not doing so, that asset tends to deteriorate. In this case the asset is the people's integrity and . . . the agent is the people themselves as politically organized" (39). Aside from its utility as a culturally unifying force, a federal government can also play a key role in mediating local disputes.

If, for example, a Native American tribe wished to build a dam in order to garner its share of Anglo-apportioned water, the national environmental lobby would find itself in a quandary. While opposing dams on principle appears noble, it seems arrogant to deny a traditionally oppressed indigenous group access to water because European-descended settlers have already festooned the arid lands with dams in a misguided attempt to turn the West into Eden. Again, the solution lies in treating the dam not as a national matter, but as a local one where voices not usually heeded enjoy unfettered access to the decision-making process. Nevertheless, the tribe's actions would inevitably impact the surrounding region and its inhabitants. Disagreements among local communities over land and resource management might best be solved by a central governing body charged with weighing the interests of all affected parties.

Currently, the tribe would have to slog through a legal system created by a bureaucracy innately hostile to its needs and beliefs. Most matters regarding the health and welfare of Native Americans are decided by the Bureau of Indian Affairs (BIA), whose primary purpose has always been the maintenance and autopoiesis of the system that conquered and confined the Indians to begin with. In the eyes of many tribes, therefore, a helpful BIA official is like a vegetarian butcher—hard to find and cause for suspicion when you do. A sustainable alternative, by contrast, would behave *as if* the Native culture, beliefs, and needs had validity equal to the scientistic, data-oriented approach that currently guides resource management decisions. This approach would not abrogate the state but reform it by making it more responsive to formerly marginalized groups.

The set of premises presently guiding the nation enforces the preeminence of a Caucasian-controlled, extractive economy predicated on limitless growth and infinite resources. While the states' rights and wise-use movements seek to continue profiting from this social system, Sustainability acknowledges limits to growth and the failure of the extractive paradigm to enable long-term survival or safeguard the health of all members of the social compact. Accepting the failure of the extractive paradigm need not involve discarding transactive capitalism, however, especially since the Marxist alternative appears equally ecologically repugnant.

The reality of capitalism in the modern era is that information has replaced durable goods as the dominant commodity in the transactive economy (see Harvey). This shift to information-based transactions has potentially revolutionary implications for economic valuation because information is not a conserved quantity. If I tell you something, or sell you a piece of knowledge, you have it, but so still do I. What I am selling you then, is *access* to the same information that I have.[18] The fundamental tenet of economic transactions—the reciprocal transfer of value—must adjust to compensate for this epistemic shift. With information itself possessing no transactive value, right of access becomes the principal commodity.

Interestingly, western water rights, with their near-total transfer of assigned value from water to water rights, seem to have anticipated the information age. Water rights and water law in general occupy a special niche in the American legal system. For example, a public right in water does not usually mean that the government owns the water. It rather means that there is a common right of use. In addition, water can never be wholly privately owned. Water rights are predicated on the notion that water is a community resource that must be shared even as the intricacies of water law acknowledge that we are all part of many different communities that form an interrelated system. Despite these egalitarian precepts, however, the western water allocation system has not functioned to the benefit of all. Communitarian ideals cannot function in a system where notions of community and access are rigidly circumscribed.

The travails within the water market also demonstrate some of the pitfalls of information trading. Water rights contain an embedded entitlement guaranteeing perpetual access to water. But, with the West's water so oversubscribed, water rights lose referentiality while suffering no corresponding loss of exchange value. In other words, even though there might be no actual water over which to hold the right, the right itself does not lose value. Allocating and trading in water that does not and may never exist is constructivism without constraint, the hydraulic equivalent of walking off a cliff without any fear of falling. No matter how many dams and diversion projects get built, the hydroeconomy will still hit the ground.

Outside the hydraulic arena, *info*trepeneurs have already adjusted to the new marketplace of ideas. Communications giants reap huge profits by offering users (*not* "consumers") access to the Internet and other informational constructs, while durable goods companies diversify, seeking to do likewise. Whether capitalism can continue to thrive amid this postmodern agora remains to be determined.[19] If so, it will necessarily mutate to accommodate the realities of what Hayles calls an "informatted society." Whatever the outcome, the information age will, without question, exert a strong influence on the national geography.

Sustainability would be a logical consequence of a cultural shift to informatics. It would behave *as if* each unity and aggregation of unities merited equal access to the unmediated flux and the cusp from which we form reality. Right now, underrepresented communities must prove that their concerns resonate with an often hostile dominant system in order to merit a hearing. Despite this gauntlet built into the system, some empowered groups advocate dismantling the few safeguards for minority access that have, over time, woven their way into the societal fabric. This agenda runs precisely counter to the compelling social need. True reform would facilitate minority access instead of deterring it. A successful state is "both a relationship of domination and an invitation to protest" (Walton 307).

In a sustainable society, if a Native American tribe wished to *dis*-mantle a dam, they would merit attention and respect instead of getting classified as Luddite enviro-extremists. Many enormous Bureau of Reclamation projects were erected without considering the needs of indigenous communities or the health of the downstream ecosystem. For example, because the Glen Canyon Dam prevents silt from washing downstream to fertilize and replenish riparian lands, Native American communities who live in the Grand Canyon and the Canyon de Chelly fight a constant losing battle to keep their land and subsistence from slipping away. Erosion has diminished both the available acreage and its yield. If it seems preposterous to propose removing these enormous structures (and in many ways it is), consider how preposterous it would have been to propose building them had a wealthy Anglo community lived where the Indians do. Furthermore, not just the lives of the various riparian communities, but the life of the river itself stand imperiled. Glen Canyon Dam's enormous storage capacity increases the depth of the reservoir's water and cools it to twenty degrees below its normal temperature. The frigidity of the water released during drawdowns kills the downstream food web.[20]

In September 1997, with many of these concerns in mind, the Sierra Club proposed to drain Lake Powell and dismantle the dam. The plan found little support among the local populace who make their livings from the $500 million tourist economy or among the Navajo, whose coal plant depends on the lake water for steam and for cooling. Nevertheless, the annual water loss to evaporation, the trapped silt that fills in the lake instead of fertilizing riparian land downstream, and the fact that the dam flooded what many call the most beautiful canyon in the world generated sufficient support to merit a hearing before Congress. The plan garnered little support there, but the fact that it got a hearing at all stands as a remarkable milestone on the way to the end of the Reclamation era.

In a parallel situation, debates rage over the dams on the Columbia and Snake Rivers that have all but destroyed their once-legendary salmon runs. Of the six hundred miles of Columbia River between the Canadian border and the Pacific Ocean, only forty-seven fall into the category of river rather than reservoir. On the Snake, the Army Corps of Engineers is considering breaching four dams that together tamed the river enough to turn Lewiston, Idaho, located five hundred miles from the Pacific coast, into an inland seaport. The Corps of Engineers' proposal has galvanized some and flabbergasted others even as it has garnered considerable congressional support. In September 1999, 107 members of Congress signed a letter to President Clinton calling the preservation of the Snake's salmon and steelhead runs "a national responsibility" and urged the administration to consider breaching the dams.[21]

The complicated regional legacy of the go-go years includes electric rates 40 percent lower than the national average and a seriously degraded ecosystem. Whether or not a radical overhaul of the network of dams in the Northwest occurs, the importance of the debate lies in the evolving social status of salmon. Despite a hostile political climate, the Endangered Species Act has forced multi-billion-dollar rescue operations aimed at preserving the remaining salmon populations. On the Snake alone, the government has already spent more than $3 billion. Whatever method is finally chosen, improving the migratory environment will likely require the spending of billions more. The Bureau of Reclamation and the Army Corps of Engineers may wind up spending billions of dollars to mimic the way the river behaved before they spent billions of dollars to alter it. Among the likely consequences: decreased power generation and lakeside recreation, and significantly higher electric rates throughout the region. It is, I think, an excellent omen that the primary beneficiary of all this spending and controversy is a fish.

As the many factions sort out the situation on the Columbia, local, state and federal officials, as well as the indigenous population elsewhere in Washington, are quietly reaching consensus over the demolition of one or two dams on the Elwha River on the Olympic Peninsula. The dams, they agree, serve little purpose, disrupt the local ecosystem, and ought to be removed. With regard to a Reclamation project west of the hundredth meridian, that type of consensus is unprecedented. Whether or not the necessary funding (recent estimates put the cost at approximately seventy million dollars) will successfully navigate the federal appropriations maze remains to be seen. As with the Glen Canyon Dam, the mere fact that billion-dollar retrofits on the Columbia or demolition of structurally sound dams on the Elwha merit serious consideration seems cause for cautious but reasoned optimism.[22] Part of the reason for these limited successes stems from the grassroots involvement rather than the intercession of national environmental organizations whose plurality of views can dilute local agendas.

Without constant vigilance, "environmentalism" slips into unproductive amorphousness. The fact that former president George H. Bush and Dave Foreman (founder of EarthFirst!) both call themselves environmentalists demonstrates that the term's meaning has attenuated past the point of usefulness. An "environmental" ethic means many things to many people precisely because the complexity of interweaving systems makes specificity impossible. In the past, large environmental groups have operated under the "big tent" philosophy. Despite sporadic successes, they have not stemmed the flow of ecologically devastating policies. If the battles were instead fought at the local level by single-cause coalitions, the effect could potentially penetrate the system's rhetorical armor. The happy result might then be a change in legitimating worldview, rather than isolated local victories.

Precedent has shown that a single-issue focus can yield significant successes. Not every self-described environmentalist supports animal rights, although overlap does occur. As a result, mainstream environmental groups have not occupied the vanguard in the movement to stop animal testing and other animal abuse. Instead, People for the Ethical Treatment of Animals (PETA) and its militant ideological ally, the Animal Liberation Front, have focused exclusively on this issue, significantly reducing animal testing while crippling the fur trade. Perhaps more portentously, animal law is now being taught at a number of major law schools.

In a different ideological arena, a similarly narrow focus has enabled antiabortion groups to balkanize the issue of choice. The platform debate at the 1996 Republican Convention testified to the potency of the single-issue approach. Antichoice forces successfully forced the party to adopt a platform calling for a constitutional amendment banning abortion, over the opposition of Bob Dole, the party's standard-bearer and despite the fact that only 6 percent of the Republican electorate supported this view.

Effective as they can be individually, when interest groups link they generate a gestalt effect. This coalescence of self-interested parties undermines the supposed inevitability of the tragedy of the commons. In Garrett Hardin's scenario, individual farmers will always maximize the number of animals they graze on a common pasture because their net gain exceeds their net loss. This conflict between self and common interest applies to virtually any environmental dilemma and, according to Hardin, is tragic because it is both foreseeable and inevitable. By comparison, linked causes avert the inevitable ruination of the commons by suggesting that farmers can find separate reasons for preserving the commons and attain consensus through self-interested reasoning.

Perhaps one farmer wants more trees in order to breed a sun-sensitive cow and another finds she can get more money per cow when they are not crowded together. Still another might not want the stream fouled because it ruins the fishing. The reasons for limiting grazing are diverse, but the net result is the preservation of the commons. This nonbinary approach to conservation demonstrates the flexibility of Sustainability's conceptual framework as well as its adaptability to an increasingly complex, information-dependent world. Even in a gigantic commons filled with competing interests that do not share a common goal, the resonance generated from each linked cause will cause the system to shift and adapt as it seeks metabalance. The result may not satisfy all constituencies, but the metabalance achieved is in the common interest, and, provided the long-term health of the system is preserved, local events can and will shift with time.

This type of compromise is less likely to occur under present conditions because certain uses of the commons command greater institutional support than others. For example, a landowner who chooses not to divert water for irrigation so as to preserve the fishing would likely lose her water right because she failed to put the water to a legally "beneficial use." The right would then go to someone who demonstrated a willingness to use the water in a state-sanctioned manner like irrigation or the watering of livestock.

The mechanistic paradigm did not rise to prominence through legislative fiat; it spidered out slowly, growing stronger as the efforts of individuals in disparate fields strengthened the trend toward objectivism. In time, the rhetoric shifted and the natural theology monolith began to give way. A similar approach could yield equally powerful results for Sustainability. Linked causes, rather than common ones, offer a version of the community-based agrarian anarchy advocated by Abbey, and in a different form, by the social ecology movement.[23]

Theories already abound as to the effects the information age will exert on the nation's geography. Some scholars, like William J. Mitchell, believe that an increasingly complex and electronically linked society will bring about a demographic "renucleation," as the need to commute to work decreases. This dwindling need for travel may cause a reemergence of small-scale urban environments, even as it increases workplace isolation.[24] Whether this scenario comes to pass or gets displaced by another, it seems logical that an increasingly linked, increasingly complex world will evolve through a coalescence of local change. Perhaps with time, the NIMBY movement can evolve into NIABY (Not in Anybody's Back Yard) and eventually to NOPE (Not on Planet Earth).

In *Learning from Las Vegas,* Robert Venturi describes taking architecture students to Las Vegas and analyzing the city from the point of view of city planning. In the students' view, the solution loomed clear: raze the city and try again. That solution would not do, so Venturi suggested that the students look at the city as a system of signs and work within the given context to mend its problems.

A similar approach could prove invaluable in restoring the ecologically ravaged western United States (including Las Vegas). The infrastructure of the Reclamation era will remain for the foreseeable future; no attempt at leveling it can wholly succeed. Nevertheless, focused, reachable goals for restoring linguistic flexibility and attaining a sustainable ecological economy may prove highly effective. Each novel surveyed in this book makes an artistic attempt to reconceptualize the status quo. The results vary. Still, a quick survey of the current status of the issues these authors grappled with offers some cause for optimism. Los Angeles, faced with courtroom defeats and a surge of negative publicity, agreed to return a significant amount of water to the Owens Valley. *The Grapes of Wrath* won valuable, if inadequate protection for migrant workers and directed the public gaze upon the untenability of the myth of the garden in the Midwest. *The Monkey Wrench Gang* spawned a new form of ecointerventionism, bathed the Glen Canyon Dam in the unforgiving glare of public scrutiny, and baldly described the social and ecological ramifications of the Reclamation era. *Animal Dreams* offers a glimpse into the continuing disempowerment of indigenous communities during the late stages of Reclamation.

All four works provide useful insights into the history of the Reclamation era and the mechanics of resistance. Hope flourishes amid the knowledge that the desert still lies beneath the Los Angeles sod and that the Colorado, if unshackled, could again run silty and thick. In short, the situation is grave, the outlook tenuous, and the rhetorical tools less than cutting edge. But the cause is not hopeless and the struggle has just begun.

Notes

Introduction

1. The hundredth meridian more or less bisects the country. It runs through the middle of North Dakota, South Dakota, and Nebraska, cutting off the western third of Kansas and Oklahoma before dividing Texas in half.

2. Ironically, the word *rival* comes from the Latin *rivalis,* which refers to people on opposite banks of a river claiming the same water rights.

Throughout this study, I make dual use of the word *reclamation.* When describing the process of water diversion and delivery, I use a small *r.* When referring to Reclamation ideology, the rhetorical basis for reclamation projects, I use a capital *R.* I use the same system when distinguishing between the practice and rhetoric of restoration.

3. Maximizing both population and available goods using a finite resource base is, as Garrett Hardin demonstrates in "The Tragedy of the Commons," mathematically impossible. This ingrained structural flaw is one of several crucial problems with Reclamation ideology and is discussed in more detail in subsequent chapters.

4. Here and throughout this study I adopt Barthes's definition of the term *myth,* using it to mean not a superstitious or erroneous belief, but rather a linguistic and cultural system through which to shape reality (*Mythologies* 113–27).

5. Francis Bacon propounded his utopian vision in *New Atlantis* (1624). In his ideal world, decisions for the good of the whole were made by scientists because they alone possessed the secrets of nature. In "Salomon's House," the community depicted in *New Atlantis,* scientists use technology to manipulate nature into a more efficient medium for production.

6. For example, Martin Lewis espouses what he calls a "Promethean environmentalism" that embraces technology and the leverage it offers in order to create a sustainable economy and habitat. Lewis maintains that, absent an international return to agrarianism, the United States cannot abdicate its leadership role in the international community or in technological

research. "Even if eco-extremists were to gain power in the United States . . . their ability to influence the evolutionary path of global society would be nil." The ensuing power vacuum would lead to the ascendancy of Japan and other nations whose environmental priorities differ markedly from those of the United States (14–15).

See also Timothy Luke, who argues that a return to an "idyllic natural past is neither likely nor necessary," and that it makes more sense to strive for new relations of production rather than seek an apocryphal premodern perfection (*Ecocritique* 203).

7. I borrow both the term *hydraulic empire* and the insight from Donald Worster's *Rivers of Empire.*

8. See also Cassuto, *Cold Running River,* an ecological biography of the Pere Marquette River in Michigan. The river's recent history shows both the boons and the pitfalls of restoring a river and then attempting to manage it in a manner that exploits its rejuvenation while maintaining its biotic integrity.

9. *Animal Dreams,* the central text of chapter 5, was published in 1990. While that is not, strictly speaking, the present, it depicts a society in the late stages of the Reclamation era attempting to craft a new sustainable paradigm. That remains our situation today.

Chapter 1

1. Americans' ignorance about the geography of the West extended to its leaders as well. For example, Thomas Jefferson, who assembled and funded Lewis and Clark's expedition, had heard that one tribe of Indians was descended from the Welsh. Part of the expedition's mission included exploring this possibility (Ambrose 154).

2. The term *wilderness* contains an inherent challenge to human civilization. *Bewilder* shares the same root and negative connotation. As Roderick Nash notes, "The image is that of a man in an alien environment where the civilization that normally orders and controls his life is absent" (*Wilderness* 2). Wilderness also implies chaos and danger, as well as, in biblical times, a lack of water. Eden, a well-watered garden and haven, could not coexist with wilderness. As Nash describes it, "The story of the Garden and its loss embedded into Western thought the idea that wilderness and paradise were both physical and spiritual opposites" (15).

3. Economist Henry Vaux points to two recurring themes characterizing the western mind-set during the first eight decades of the twentieth century. The first was that water's scarcity should not constrain economic growth in the region. This was the fundamental principle and justification for the ongoing development of the region's scarce water. The second principle involved

what he calls "preemptive development," by which he meant people had to develop and use the water before someone else did (Sax, Abrams, and Thompson 682).

4. The theory that rain followed the plow drew its scientific substantiation from, among other sources, the theories of James Pollard Espy, one of the pioneers of meteorology. Espy believed that as the frontier expanded westward, the clearing of the land actually created a different weather system, warming the region and engendering precipitation for the Plains. This theory dovetailed nicely with the belief that the "'national weather' was, in effect, exactly coterminous with the ever-expanding boundary of the nation's body" (Ross 224).

5. Though his pronouncements on the climate and geography of the West drew on the writings of such nineteenth-century luminaries as Darwin and Humboldt, Gilpin's theories were his own and lacked the rigor and discipline of his eminent antecedents. That lack of rigor was a crucial failing; it led to conclusions that were "*a priori,* deduced, generalized, falsely systematized, and therefore wrong" (Bernard DeVoto, qtd. in Emmons 9).

6. For a fuller discussion of the common interests and subsequent cooperation between the federal government and the railroads, see H. N. Smith, especially chapter 2, and Emmons.

7. The sexual connotations of this term and its accompanying belief system were not lost on those men who first laid claim to the land, nor on the historians who later analyzed their actions. *Virgin Land* is the title of Henry Nash Smith's pioneering study of the mythology of the American West, answered by Annette Kolodny's ironic title, *The Lay of the Land.* The progression of the land from earth mother/maiden goddess to conjugal partner/slut has been the subject of several excellent analyses in addition to Smith's and Kolodny's, among them the works of Susan Griffin.

8. Because equipment loss prevented the accumulation of much important data, Powell led another expedition down the Colorado in 1872. His published account of those trips, *The Exploration of the Colorado River and Its Canyons,* merges the two journeys into one, primarily for literary purposes.

9. For a more encompassing view of Powell's journey, see his journals. For a more complete look at the man himself, as well as the historical climate in which he worked and lived, see Wallace Stegner's excellent biography, *Beyond the Hundredth Meridian,* as well as *The American West as Living Space.*

10. An acre-foot is 325,850 gallons, enough water to cover one acre of land to a depth of one foot. One acre-foot contains enough water to supply three average homes for one year.

11. According to Stegner, Powell's ideas ranged "so far beyond the social

and economic thinking of the period" that they "seem like the product of another land and another people" (*Meridian* 228).

12. For comparison's sake, New Mexico alone is almost three times the size of Ohio.

13. Here and throughout this book, my use of the terms *signifier, signified,* and *sign* follow Saussure's classifications wherein *signified* corresponds to use value, *signifier* relates to exchange value, and *sign* unites the two.

14. The reasonable-use doctrine was codified and explained in the 1827 Rhode Island court case, *Tyler v. Wilkinson.*

15. Beneficial use is a legal term of art and its definition varies from state to state. All western states, however, classify domestic, municipal, agricultural, and industrial uses as beneficial. In recent years, many have also included recreational uses. Once a user puts water to a use deemed beneficial under state law, the water right is perfected. That means the user's right to that water cannot be defeated, even in times of shortage, by other uses that might have more social value but which were asserted after that first beneficial use. Thus, in the West a farmer growing rice in the desert can prevent a city from receiving water for its residents.

16. For a good summary of western water law, as well as a comparison with riparianism, see Wehmhoefer. See also Getches.

17. The Desert Lands Act was more than merely the statutory origin of the rule of prior appropriation; it was also an incarnation of the Homestead Act. It stipulated that settlers must show "proof of irrigation" before taking title to a plot of land. Unless the land bordered a river that did not flow through a canyon (unlike most rivers in the arid region), irrigation became enormously expensive—too expensive for the average settler. The result was rampant speculation, water monopolies, and fraud.

18. John Muir had tried to construct a nonanthropocentric, "nature for nature's sake" vocabulary when he led the fight to prevent the damming of the Hetch-Hetchy Valley from 1906 to 1913. His efforts failed principally because he could not communicate his message in a manner that captured the imagination of the common American. His rhetoric was rich in spirituality and the aesthetics of the sublime. Those who loved the outdoors as Muir did readily embraced his rhapsodic descriptions, but a nation of average Americans caught up in the enthusiasm of the progressive era could not relate to it.

19. Leopold's land ethic, published in 1949 as part of *A Sand County Almanac,* has become one of the foundational principles of conservation and enormously important to the growing discipline of ecophilosophy. See, for example, Callicott.

Chapter 2

1. The term *resource,* or to "rise again," originally referred to the earth's ability to restore itself. People used resources commensurate with nature's ability to rejuvenate and replace what was lost. By the nineteenth century, however, ecosystemic needs no longer entered the equation. In *Natural History of Commerce* (1870), John Yeats wrote, "In speaking of the natural resources of any country, we refer to the ore in the mine, the stone unquarried, the timber unfelled, etc." (qtd. in Merchant, *Revolutions* 11). The result, in Annette Kolodny's view, was "the transforming of nature into wealth" (*Lay* 133).

2. The *Los Angeles Times* trumpeted Lippincott's actions as those of a hero, ignoring their obvious impropriety and focusing instead on the city's windfall: "Without Mr. Lippincott's interest and cooperation, it is declared that the plan never would have gone through. . . . Guided by the spirit of the Reclamation Act . . . he recognized that the Owens River would fulfill a greater mission in Los Angeles than if it were spread over acres of desert land. . . . Any other government engineer . . . undoubtedly would have gone ahead with nothing more than the mere reclamation of the arid lands in view" (qtd. in Reisner 70). The irony of Lippincott's supposedly being "guided by the spirit of the Reclamation Act" while actually sabotaging a potential reclamation district and destroying the agricultural capacity of a river and valley was not lost on valley residents. To those in the city, though, who saw both the Reclamation Act and the city's actions as part and parcel of a progressive agenda, no irony existed. Because the majority would benefit, the spirit of the act had been obeyed.

3. City spokesmen did not speak of future needs, however. Instead, they issued dire proclamations trumpeting the current "water famine." William Mulholland, superintendent of the Los Angeles water system and architect of the plan to divert the Owens River, did everything he could to instill the project with a sense of a terrible urgency, claiming that the city had been experiencing a severe drought since the mid-1890s. This claim may be overstated—rainfall had averaged less than three inches below normal for the years 1900–1905 (Hundley, *Thirst* 150).

4. Donald Worster notes this flaw in Powell's reasoning, observing that, while visionary, Powell's ideas offered no solution to the West's land-use dilemmas: "If there was one factor that would defeat broadly diffused democratic tenure in the region, it was precisely the demand for ever higher living standards. . . . Nothing in Powell's proposals would have stopped that process, for he had not begun to question the underlying economic values of the culture" (*Dust Bowl* 87).

5. In a strict sense, the city's actions were within the law. "Los Angeles employed chicanery, subterfuge, spies, bribery, a campaign of divide and conquer, and a strategy of lies to get the water it needed. In the end, it milked the valley bone-dry, impoverishing it, while the water made a number of prominent Los Angeleans very, very rich. There are those who would argue that if all this was legal, then something is the matter with the law" (Reisner 65).

6. The Los Angeles Board of Public Service Commissioners blamed the valley's resistance to the aqueduct on "the mental reactions of a pioneer community . . . uninformed and unaccustomed to the ways of the outside world" (qtd. in Walton 184).

7. These acts included several occupations of the pumping station (one of which turned into a weeklong town picnic) and a number of dynamite blasts to the aqueduct.

8. The *Los Angeles Herald* bragged in 1905 that the Owens River's annexation would ensure that "lawns . . . could be kept perennially as green as emerald and greater Los Angeles could go on swimmingly in its metropolitan progress" (qtd. in Hundley, *Thirst* 150). The editor's choice of the adjective "swimmingly" is particularly apt in light of the city's soon-to-be-realized status as swimming pool capital of the world.

9. As Walton notes, "The tradition-inspired revolt in the Owens Valley took place within the framework of state modernization and economic growth, not as a backward-looking attempt to restore some pre-capitalist Eden or resist the penetration of market forces. On the contrary, it sought fulfillment of the developmental promises that lured pioneers and spoke to western progressives" (194). The revolt in the valley appears less inspired by yeoman values than by a desire to modernize the region's agricultural apparatus.

10. I have provided a rudimentary outline of what was and continues to be a very complex, highly divisive issue in California's history and in national water policy. For a more thorough treatment of the regional history and the causative factors involved, see Walton; Hundley, *Thirst;* and Reisner. There are also a number of fictional treatments of the valley's struggles (other than *The Ford*), most of them melodramas, including Frances Gragg and William Putnam's *Golden Valley* (1950) and Cedric Belfrage's *The Promised Land* (1938).

11. Richard Hofstadter notes that the tendency to treat the land as a commodity rather than as a source of subsistence gradually changed the western farmer from yeoman to businessman: "The characteristic product of American rural society was not a yeoman or a villager, but a harassed little country businessman who worked very hard . . . gambled with his land, and made his way alone" (46). That transition from farmer to businessman contained the

seeds of ecological destruction for the arid western lands. A region without abundant rainfall is far more ecologically fragile than a humid zone. Factory farming, involving large-scale monoculture and intensive irrigation, destroys the soil's nutrient balance while salinating what little water the region retains. These phenomena become more apparent as the Reclamation era progresses and are discussed more thoroughly in later chapters.

12. Stephen Jay Gould has long been one of the most vocal and perceptive critics of the ruse of scientific objectivity (see *Mismeasure of Man,* especially pp. 19–29). See also Alan Gross's discussion of the need to recast scientific advances as inventions rather than discoveries. To discover, Gross argues, is to find out what already is there. Science can make no such claim, as the ongoing superannuation of scientific theories makes plain. Invention, on the other hand, "captures the historically contingent and radically uncertain character of all scientific claims, even the most successful" (7). If we treat scientific theories as rhetorical inventions, the need to explain their vulnerability to obsolescence disappears.

13. This insight is taken from Annette Kolodny's excellent discussion of the novel as illustrator of American terrestrial values (*Lay* 138).

14. Susan Griffin describes land's shifting identity: "He has made her conceive. His land is a mother . . . Again and again, in his hunger, he returns to her . . . She is his mother. Her powers are a mystery to him. Silently she works miracles for him. Yet, just as silently, she withholds from him. Without reason, she refuses to yield. She is fickle. She dries up . . . He is determined he will master her. He will make her produce at will. He will devise ways to plant what he wants in her, to make her yield more to him" (53).

15. Environmentalism at the turn of the century was known as "conservationism" and was riven with internal conflicts between "conservationists" and "preservationists." Conservationists favored judicious development of land and resources, while preservationists rejected utilitarianism in favor of safeguarding an unspoiled nature. The preservationist cause was championed by John Muir and botanist Charles Sprague Sargent among others, while the conservationists were led by Gifford Pinchot and Frederick Newell (Nash, *Wilderness* 129–40).

Feminism during Austin's time described a political position more than a philosophy. Feminists were women seeking equal rights, including the right to vote. The philosophical tenets of feminism have been around for a long time. Historians traditionally located the origin of "feminist consciousness" in the nineteenth century, coinciding with the nascency of the political women's rights movement. However, some modern historians trace it as far back as the fifteenth century and the works of French author Christine de Pizan (see Lerner 14–15).

16. The Austins were separated by the time *The Ford* was written.

17. This phenomenon crops up again in later years as those fleeing the Dust Bowl curse the banks and factory farms that drove them from their land even as they themselves rushed to adapt the same ecologically disastrous land-use strategies to their smaller farms. It was these methods more than any shortfall of water that created the Dust Bowl. See chapter 3.

18. One need only look at Texas, with its cities and communities built on oil wealth, and the fortunes of the state integrally linked to the world petroleum market, to see the tremendous impact oil can have on a region. Even with the collapse of the domestic oil market, the economy of Texas, as well as of Louisiana, Oklahoma, and several other states, remains tied to the petroleum industry.

19. Karen Warren and Jim Cheney define ecofeminism as a movement that "attempts to unite the demands of the women's movement with those of the ecological movement in order to bring about a world and worldview that are not based on conceptual structures of domination" (179). They go on to describe a methodology based on "observation set theory" that posits numerous realities while privileging none. Anne seems to be alluding to a similar concept in her suggesting that the dominance of androcentric culture lies in "the way we look at things."

20. Andrée Collard argues in *The Rape of the Wild* that female experience shares a history of oppression and abuse with nature, thus making women more sensitive to such actions and better equipped to remedy them (138). The self-imposed alienation of men from nature is inherently damaging to the biotic regime: "A culture that defines adulthood in terms of one's ability to separate from mother/nature and defines mental health according to the smoothness of this separation, is a culture that denies its life blood. Such a relationship is bound to be destructive" (28).

21. This is by no means the only view advanced by ecofeminists. While Andrée Collard, Dorothy Dinnerstein, Susan Griffin, and others advance this view, many do not. Catherine Roach, for example, rejects the notion of a closer link between women and nature, arguing persuasively that constructions such as women/nature and men/culture perpetuate a patriarchal, dualistic worldview. They also syllogistically separate nature and culture into two discrete entities. Roach believes that this alienation lies at the heart of the current ecological crisis: "This dualism is unsound because it encourages the belief that 'culture' and humanity are quite apart from 'nature' and that we humans may thus use and abuse the environment at will, without ourselves suffering from the damage we inflict. Any understanding of the world that posits an important or unbridgeable difference between the realm of the human and the nonhuman risks creating a gulf between the two in which the human, because of our inherent chauvinism or anthropocentrism, would inevitably be more valued than the nonhuman" (54). I offer this capsule view

of the diversity within the ecofeminist community in order to dissolve any boundaries I may have inadvertently drawn around the doctrine of ecofeminism. It is not my intent either in this note or in this study to do justice to the richness of ecofeminist discourse. I merely wish to point out the relevance of Mary Austin's philosophy to the ecofeminist vision that celebrates the women/nature relation.

Chapter 3

1. I am indebted to Roderick Nash for the biblical references, originally made in *Wilderness and the American Mind,* 14.

2. The trend was visible as early as 1847 to George Perkins Marsh, one of the first and most influential American ecologists and an early opponent of the myth of the garden. In a speech to the Rutland County Agricultural Society he stated that in the United States "the full energies of advanced European civilization, stimulated by its artificial wants and guided by its accumulated intelligence, were brought to bear at once on a desert continent" (qtd. in Marx, *Machine* 204).

3. Such is technology's domination over the agricultural process that growers in California's Imperial Valley do not even like it when it rains; the sun soon overheats the dampened crops and wilts the lettuce leaves (Reisner 4).

4. This tendency to privilege water, either by scarcity or surfeit, appears frequently in the Steinbeck canon. For example, *Of Mice and Men* opens and closes on the banks of a river; *The Log from the Sea of Cortez,* with its fascination with tide pools, offers the clearest presentation of Steinbeck's ecophilosophy; and *The Wayward Bus,* like *The Grapes of Wrath,* utilizes floodwaters to spur its characters to action and the acquisition of wisdom.

5. Congressman Lyle Borden of California declared *The Grapes of Wrath* to be "a lie, a black, infernal creation of a twisted, distorted mind" (qtd. in Steinbeck, *Working* xxiv). Steinbeck became the target of a whispering campaign by the growers' alliance, the Associated Farmers, one rumor being that Steinbeck was a Jew acting on behalf of a Zionist-Communist conspiracy to undermine the economy (Benson, *Adventures* 420).

6. The Associated Farmers believed the publications of *The Grapes of Wrath* and *Factories in the Field* to be evidence of a vast left-wing conspiracy, the supposed existence of which fueled their outrage at both authors. Prior to their publication in 1939, the governor of California, Culbert Olson, had also made McWilliams chief of the state's Division of Immigration and Housing. The job included enforcing the long-dormant state labor camp act. McWilliams's appointment further stoked the growers' rage and paranoia.

They labeled him "Agricultural Pest No. 1 in California, outranking pear blight and boll weevil" (McWilliams x).

7. Adherence to the doctrine of manifest destiny, even before it was so named, allowed white settlers to casually displace Native Americans without regard for history, negotiated treaty, or first right of occupancy. Completely dismissing Native claims to their homelands permitted the notion of "virgin land" to arise. Acknowledging Native claims meant dismantling an intact and seductive mythology as well as relinquishing the American right not just to continue expansion, but to exist at all.

8. That is not to say that there was not a serious drought in the Plains during the 1930s. In fact, almost the entire nation endured a terrible drought from roughly 1930 to 1936 (Hoyt 8–9, 66).

9. James N. Gregory argues that the dust storms displaced fewer people than government policies designed to bolster falling crop prices. In 1933 the federal government passed the Agricultural Adjustment Act, which paid subsidies to landowners who took their land out of production. Since the majority of farms in the Plains were operated by tenants (like the Joads), taking the land out of production effectively deprived them of their livelihood. The act compensated landowners for not planting, but left tenant farmers without food or means of earning a living (11–13).

10. Muley's unbreakable bond with his homeland offers a powerful parallel to the link between many oral cultures and the land from which their culture draws both sustenance and definition. There is, according to philosopher-ecologist David Abram, a powerful link between nonwriting cultures (many Okies, including most of the Joads and Muley, were barely literate) and the places in which their stories take place. "A particular place . . . is never, for an oral culture, just a passive and inert setting for the human events that occur there. *It is an active participant in those occurrences.* . . . [T]he place may even be felt to be the source, the primary power that expresses itself through the various events that unfold there" (162).

11. Though it does not directly treat reclamation, *The Education of Henry Adams* poignantly encapsulates the growing pains of American society as it struggled to cope with an expanded technological capacity and the new challenges presented by the twentieth century.

12. One of the fascinating aspects of this passage is that Joseph's experience is portrayed less as autoeroticism than as a sexual encounter between a man and the earth: "[T]he exultation grew to be a sharp pain of desire that ran through his body in a hot river. He flung himself face downward on the grass and pressed his cheek against the wet stems. . . . His thighs beat heavily on the earth" (*God* 11). There is no element of fantasy; he does not imagine that he is having sex with another person. He is mating with the land.

13. Kolodny argues that the progressive deterioration in cultural rever-

ence for the land was an unavoidable by-product of viewing it as feminine while seeking to settle it: "Implicit in the metaphor of the land-as-woman was both the regressive pull of maternal containment *and* the seductive invitation to sexual assertion: if the Mother demands passivity, and threatens regression, the Virgin apparently invites sexual assertion and awaits impregnation" (*Lay* 67).

14. Worster offers this account of the Plains mentality during the mid-1930s:

> "You gave us beer," they told Roosevelt, "now give us water." . . . "Every draw, arryo [*sic*], and canyon that could be turned into a lake or lagoon," wrote a clothing store manager, "should be turned into one by dams and directed ditches & draws until there are millions of them thru these mid-western states." A Texas stockman wanted to use natural gas to pump flood waters from the Mississippi River to the Plains. . . . An old soldier from Denver penciled his ideas on ruled tablet paper: stage sham battles with 40,000 Civilian Conservation Corp boys and $20 million worth of ammunition—the noise would be sure to stir up some rain. . . . "Try it," he finished, "if it works send me a check for $5000 for services rendered." (*Dust Bowl* 39)

15. Gregory maintains that the westward migration was due less to any organized effort by the growers than by "migration chains" whereby migrants who had moved west during better times (i.e., the 1920s) and found work wrote their relatives about their success. When dust storms and the depression hit the Plains, afflicted families remembered these success stories and sought to emulate them. Hopeful songs trumpeted the splendor they hoped to find upon their arrival in the Golden State:

> California, California,
> Here I come too.
> . . . .
> Nothing's left in Oklahoma,
> For us to eat or do.
> And if apples, nuts, and oranges
> And Santy Claus is real,
> Come on to California,
> Eat and eat till you're full.

The California government, concerned that such rosy-hued rumors would cause the state to be overwhelmed with refugees, made some token efforts to discourage migration. These efforts included posting a billboard on Route 66 outside of Tulsa that read:

No JOBS in California
IF YOU are looking for work—KEEP OUT
6 Men for Every Job
No State Relief Available for Non-Residents.

The tourist industry was also concerned about the potential effects of mass migration on the state's allure. Vacation ads featured fine print warnings like "Come to California for a glorious vacation. *Advise anyone not to come seeking employment, lest he be disappointed;* but for tourists, the attractions are unlimited" (qtd. in Gregory 22).

16. Steinbeck had already devoted an entire novel *(In Dubious Battle)* to the political implications of the worsening tensions between growers and migrants. In *The Grapes of Wrath,* he ontologizes the migrant struggle and makes the inevitability of social change even more evident. Both novels embody his "phalanx" theory, which held that the collective will of the people differs from the sum of its component parts; it is a unique entity whose force far exceeds that of its members.

17. According to one California agricultural spokesman: "We are not husbandmen; we are not farmers. We are producing a product to sell" (qtd. in Worster, *Dust Bowl* 57).

18. Foucault believes that approaching knowledge through the study of geography ("region, domain, implantation, displacement and transposition") leads to true awareness of the relationship between power and knowledge (*Power/Knowledge* 69).

19. Marx's description of worker alienation (in "Wages of Labor") is uncannily accurate when related to the migrants: "Since the worker has sunk to the level of a machine, he can be confronted by the machine as a competitor. Finally, as the amassing of capital increases the amount of industry and therefore the number of workers, it causes the same amount of industry to manufacture a *greater amount of product,* which leads to overproduction and thus either ends by throwing a large section of workers out of work or by reducing their wages to the most miserable minimum" (69). The first part of the quotation could easily be describing the situation in Oklahoma, while the second half diagrams the Okies' dilemma in California.

20. Dehumanizing other people is a complicated process. It is "fraught with deep ambivalence and anxiety" and stems from the special value we attach to human life. Groups try to objectify other groups even while admitting that the "objects" are in fact human. The victims become, in the dominators' eyes, not quite human and not quite object (L. Cassuto chap. 1).

21. After visiting a series of migrant camps in 1940, Mrs. Roosevelt told reporters, "I have never believed *The Grapes of Wrath* was exaggerated" (qtd. in Benson, *Adventures* 402).

22. In March 1950, the *New York Times* reported, "The spectre of human misery again is striking the Grapes of Wrath country. Throughout the vast and fertile San Joaquin valley . . . where the trek of the Okies made history a dozen years ago, a new cycle of destitution among farm workers is under way" (qtd. in McWilliams xii).

23. Robert Reinhold, "U.S. Says Scarce Water Supplies Won't Go to California Farmers," *New York Times,* February 15, 1992, 1+.

Chapter 4

1. Ecotage, or "monkeywrenching" (so named in honor of the novel) is, according to Dave Foreman, "non-violent resistance to the destruction of natural diversity and wilderness . . . aimed at inanimate machines and tools that are destroying life. Care is always taken to minimize any threat to people, including the monkeywrenchers themselves" (113). Abbey clarifies the term in his preface to the monkeywrencher's handbook, *Ecodefense:* "[I]f the wilderness is our true home and if it is threatened with invasion, pillage and destruction—as it certainly is—then we have the right to defend that home . . . by whatever means are necessary" (qtd. in Foreman 141).

2. Potlatch is still practiced by some Northwest tribes, but in a much altered form. It all but died out due to suppression from provincial authorities and efforts by the Canadian and United States governments, as well as by many Indians, to assimilate the Indians into the dominant culture.

3. "The hosts at one potlatch were the guests at another and in this sense potlatches are reciprocal. . . . Thus, there is an endless chain of potlatching and distributions of property between groups" (Rosman and Rubel 29).

4. For a fuller discussion of the potlatch's demise, see Cole and Chaikin's *An Iron Hand upon the People.* Ironically, contact by European settlers propelled the potlatch into far greater prominence among the tribes than it had previously enjoyed. Wealth acquired through the fur trade and later through commercial fishing and logging allowed Indians to spend far more lavishly than they could before. Precontact potlatches tended to be small affairs with handmade blankets and crafts serving as the principal gifts. Postcontact potlatches evolved into hugely expensive gatherings with commercial gifts often carrying the most value. It was only after contact that potlatches began to genuinely impoverish their participants. In this sense, capitalism sowed the seeds for the potlatch's demise before religious zeal and governmental litigiousness even entered the fray.

5. "In Bataille's view, the bourgeois individuals . . . who foster a desire to revolt by soaring 'above' are destined for a fall, and in a way, *want to fall:* thus the 'Icarian Complex,' an 'unconscious' and pathological desire to fall. Icar-

ian revolt . . . is the only pathology Bataille will condemn; it is the pathological refusal to embrace stinking decomposition" (Stoekl, in Bataille xv; emphasis added).

6. Andre Gorz integrates these arguments into an ecophilosophical context. In advocating bioregionally independent communities similar to those envisioned by Abbey for his agrarian anarchist society, Gorz's rhetoric overtly recalls Bataille: "All production is also destruction. This fact can be overlooked so long as production does not irreversibly deplete natural resources. . . . They regenerate themselves naturally—the grass grows back along with the weeds. The effects of destruction appear wholly productive. More precisely: this destruction is the very condition of production. It has to be repeated again and again" (20).

7. Approaches to environmental management in the United States traditionally assume opposition as on a balance sheet. Environmental impact statements are, by their very nature, procedures by which environmental degradation is weighed against proposed social gain. Andrew Ross notes that "the game of winning general consent for ideas in the history of modern capitalism has been waged on a balanced terrain of contradictions, where narratives about traditional values are played off against narratives of progress. Something always has to be sacrificed for progress to be made" (232).

8. Abbey recounts with pleasure Jeffers's famous advice, "Be in nothing so moderate / as in love of man" (qtd. in *Desert* 177).

9. Hofstadter observes that it is "too little realized that the farmers, who were quite impotent as a special interest when they were numerous, competing and unorganized, grew stronger as they grew relatively fewer, became more concerted, more tenaciously organized and self-centered" (7).

10. Leo Marx argues that Thomas Jefferson among others envisioned the machine as the agent with which to maximize the land's productive potential: "Once the machine is removed from the dark, crowded, grimy cities of Europe . . . it will blend harmoniously into the open countryside of his native land . . . turning millwheels, moving ships up rivers, and . . . helping to transform a wilderness into a society of the middle landscape" (*Machine* 150).

11. Phillip Fradkin points out that "cows are far and away the chief beneficiaries of water, even in the West's most populous region. . . . The Imperial Valley [of California] was the single largest user of Colorado River water; the most valuable single crop . . . was alfalfa, used for hay" (33). This is a telling statistic. The metropolitan areas of Los Angeles, Phoenix, Las Vegas, and so on, do not approach the consumption levels of corporate farms. Nonetheless, to say that cows are the beneficiaries of this water is misleading. Cows are grazed, fattened, and slaughtered to benefit the ledgers of agribusiness. They are no more responsible for their presence than the river is for its role in irrigation.

12. In 1975, the year *The Monkey Wrench Gang* was published, 2.3 million acre-feet evaporated from Colorado River reservoirs. That amount exceeded the total water consumption of the four upper-basin states and Nevada (Fradkin 76).

13. In 1999 dollars, the cost exceeds one billion.

14. Bonnie's relatively ancillary role in the gang's activities as well as her status as object of desire and disruptive force buttress oft-made charges of misogyny against Abbey. These charges, while perhaps justified and certainly deserving of further discussion, do not fall within the purview of this study.

15. Bishop Dudley Love is an interesting character in his own right. In many ways he is the antithesis of both Doc and Smith. Like Doc, he is wealthy and community-oriented. Yet he uses his wealth to develop the canyon lands and to fuel his own political fortunes. Love's community service consists of heading the Search and Rescue Team, a group of volunteer law-enforcers who hunt down the monkeywrenchers. And, like Smith, Love is a Mormon whose behavior does not fit with traditional standards of piety. Though a bishop in the Mormon Church, Love's religiosity is driven by cynicism and greed. As Smith tells it: "We got plenty like him in Utah. . . . They run things as best they can for God and Jesus, and what them two don't want, why fellas like Bishop Love pick up. They say it's a mighty convenient arrangement all around. Jesus Saves at eight and a half percent compounded daily" (*Gang* 109–10).

16. In Abbey's novel *Good News* (1980), Burns is the hero of a dystopian science fiction novel set in a postnuclear Southwest. Just as in *The Monkey Wrench Gang* and *Hayduke Lives* (1989), Burns is a ghostly figure from the past who rides into a skewed present and attempts to restore reverence for the land.

17. Paul Horgan, himself an accomplished writer of the West, wrote in 1964, "If there is a single pervasive theme in writing about the *west,* . . . it could be identified as the theme of man, alone, against the grand immensity of nature—the nature of the land, reflected in his own soul" ("Western Novel" 28).

18. Abbey's formulation offers an interesting parallel to John Walton's discussion of the relationship between state and culture in *Western Times and Water Wars.* Walton, paraphrasing Phillip Corrigan and Derek Sayer, notes that "state forms are cultural and cultural forms are state-regulated. The state is not an object but an 'exercise in legitimation [and] a bid to elicit support for' domination which succeeds to the extent that it constructs cultural justification" (307). For both Abbey and Walton, culture is not the celebrated product of societal advance but rather a dead weight on social evolution, an obstacle to be overcome.

19. For a concise encapsulation of Abbey's philosophy of "agrarian anar-

chy," see his essay "Theory of Anarchy" (*One Life* 25–28). For a more theoretical discussion of ecoanarchist bioregionalism see Murray Bookchin's *Remaking Society,* particularly 185–204. Abbey and Bookchin differ on many key ecological and ideological concepts, but both endorse utopian, bioregionalist visions wherein technology functions within the limits of a given ecosystem.

20. Richard Grusin maintains that the belief in the beneficialness of nature's profligacy owes much to Thoreau's formulation of nature's economy in terms of symbolic rather than market value. He points to "Wild Apples" and "Autumnal Tints" as clear indicators of Thoreau's reconfiguration of the natural economy: "Each essay takes as its subject natural objects that society thinks of as useless . . . and proceeds to show that both are invaluable. Thoreau aims to point out that his contemporaries misvalue nature not because they insist on seeing it in economic terms but because they insist on understanding its economy in terms of market exchange rather than symbolic expenditure" (38). Nature's profligacy, according to Thoreau, is something to emulate rather than criticize. In his journals, he praises the system that "can afford to let so many moons burn all the day as well as the night—though no man stands in need of the light. . . . Nature would not appear so rich . . . if we knew a use for everything. . . . Her motive is not economy but satisfaction" (qtd. in Grusin 36). This Thoreauvian system of natural theology stood in stark contrast to the technocentric paradigm that was fast taking hold of the nation.

21. I do not claim that Abbey attempted to integrate Bataille's formulations into his fiction or that Abbey had even read Bataille. However, Abbey did hold an M.A. in philosophy from the University of New Mexico, and his oeuvre suggests a familiarity with a number of philosophical schools. Furthermore, the fact that *The Monkey Wrench Gang* is dedicated to Ned Ludd, godfather of the Luddite movement, certainly indicates Abbey's sympathy for such a view.

22. In *Desert Solitaire,* Abbey quotes Bakunin to make a similar point. Bakunin states, "There are times when creation can be achieved only through destruction. The urge to destroy is then a creative urge" (*Desert* 185).

Chapter 5

1. Biologist Francisco Varela defines an autopoietic system as one that is both autonomous and continually self-producing (13). In *Ecological Communication,* Niklas Luhmann adapts Varela's concept to social systems, arguing that when system elements are conceived of as communicative acts

rather than bioenergetic entities, the concept of autopoiesis extends to the social domain. For an excellent summary of the way the biological notion of autopoiesis maps onto social systems, see Paulsen 121–27.

2. "Unlike the concept of meaning, that of information is always to be understood relative to an actually given, constantly changing state of knowledge and individually structured preparedness to process information. The very same meaning complex can thus result in quite different information, depending on when and by whom it was actualized in experience" (Luhmann, *Essays* 31). Or, to adapt William Paulsen's analogy, one person's lucid intelligent conversation on one phone line becomes mere random noise when it bleeds into another conversation on another line (67).

3. Paulsen argues that language is itself both an autopoietic system, "one that has defined and continues to define itself in the course of the recursive interactions that are its use," and also an allonomous system, one that serves a purpose outside itself (129–30). This characterization is very close to the idea of language as a function system in the maintenance of the larger unity. It does, however, help explain language's centrality within each function system that makes up the unity. It further illuminates why language is sometimes ineffective as the communicative medium for the unity: "[I]t has its own internal self-referential laws . . . the features of an autonomous system that is the accretion of all the other messages that have been sent within it" (130).

4. Luhmann observes that "evolution is only possible on the basis of deviant reproduction of systems" (*Theory* xii). For a good discussion of complexity as a beneficial adaptive mechanism, see La Porte.

5. This contrasts with the mechanistic worldview that "arose simultaneously and in support of early capitalism. . . . Mechanism and its ethic of domination legitimates the use of nature as commodity, a central tenet of industrial capitalism" (Merchant, *Radical* 11).

6. Tailings are the waste product created by the mining and refining process. They often contain heavy metals and other toxic materials.

7. For example, the Homestead Act's requirement that settlers display an "erected domicile" on their land to show that they meant to live on the property created a boom in the sales of birdhouses. Speculators placed the birdhouses on their property in order to comply with the letter, if not the spirit of the law.

Another famous example of quasi-legal fraud involves Henry Miller's acquisition of his 1,090,000-acre ranch in California's Central Valley. The Swamp and Overflow Act stated that any federal land that overflowed enough to allow traversing in a flat-bottomed boat could belong to anyone who promised to drain and reclaim it. According to legend, Miller bought a boat, hitched it to a horse-drawn wagon, and dry-land boated across several counties (Reisner 46).

8. For variations on this same theme that specifically treat native New Mexicans, John Nichols's *The Milagro Beanfield War* (1974) provides a fictional account, while Stanley Crawford's *Mayordomo* (1986) offers a compelling first-person narrative of the author's experiences as ditch boss on an *acequia* (irrigation ditch).

9. Ivan Illich describes the resulting schism between traditional and contemporary visions of water: "H_2O and water have become opposites. . . . H_2O is a social creation of modern times, a resource that is scarce and that calls for technical management. It is an observed fluid that has lost the ability to mirror the water of dreams" (76).

10. Politics is the shifting arena in which contested processes inevitably land. When function systems cannot agree on who retains regulatory authority for a given process, the arbitration of the dispute is political: "The third value that is excluded from the code of legal and illegal, i.e., what for the time being is neither legal nor illegal, appears in the legal system as politics. Thus the legal legitimization of political decision-making leads to the reintroduction of the excluded third value into the system" (Luhmann, *Communication* 73).

11. There are many such examples. The most topical is perhaps that of the Mexican migrant workers in California. Prior to the arrival of the exodusters, Mexican migrant workers had been brutally exploited with little thought given to their welfare or even their survival. The growers justified their treatment of the Mexicans by reasoning that the Mexicans were just "living as they normally do, making themselves available to serve whites in their quest to make the desert a garden" (Mitchell 109).

In *White by Law,* Ian Haney-López describes the strained reasoning process used to determine "legal" whiteness for purposes of American citizenship. Racial classifications for immigration served the dual purpose of defining the nation's borders and the borders between races. Haney-López uses several landmark court cases from the first two decades of the twentieth century to demonstrate the value of whiteness to whites and the accompanying need to define it in such a way as to further rarify the status of white people.

For another discussion of dehumanization through boundary drawing, see Andrew Light's essay "Urban Wilderness." Light argues that in a world where the frontier has long since dissolved, the heart of darkness now beats in the inner city, a place where middle-class fears combine with racism and difficult living conditions to create an urban "jungle" of the type first depicted by Upton Sinclair.

Light points to the film *Falling Down,* where an L.A. defense worker played by Michael Douglas decides he cannot take it anymore and attempts to clear out the jungle by force. The filmmakers claimed the film was a critique of racism and simplistic attitudes about the inner city. Unfortunately, film audiences saw it instead as an accurate portrayal of the urban wilderness, and audiences identified with its portrayal of a righteous inner-city crusader. In Light's view, absent a prevailing notion that the inner city is the only truly uncivilized place remaining, films like *Falling Down* would not be possible.

12. Luhmann comments that "systems define their own boundaries. They differentiate themselves and thereby constitute the environment as whatever lies outside the boundary" (*Communication* 6). It is worth noting that the act of drawing boundaries and defining environment is a self-interested act. The system will designate boundaries conducive to its autopoietic potential.

13. Mary Douglas observes that "pollution behavior is the reaction which condemns any object or idea likely to confuse or contradict cherished classifications" (36).

14. The Trail Smelter Arbitration has been the subject of a great deal of scholarly interest both in and outside of the international legal community. For a history of the smelter and the region, see Turnbull. For concise and readable discussion of the legal side of the smelter's history, see Rubin.

15. An ironic coda to the Trail Smelter story further interrogates the concept of damage. Since 1906, the Trail Smelter has been owned by the Cominco Corporation, a multinational mining concern with mines and interests all over the world. Some of Cominco's holdings are in Madagascar, where mining companies are currently viewed by some as ecofriendly because the mines cause less harm to the local environment than the rampant deforestation taking place elsewhere on the island.

16. See Mary Douglas's discussion of dirt and pollution as windows through which to view a system's ordering methods. Dirt, according to Douglas, is never an isolated event. "Where there is dirt there is a system" (35).

17. "This linkage of human to human is, in the final analysis, the groundwork of all ethics as a reflection on the legitimacy of the presence of others" (Maturana and Varela, *Tree* 247). A group can, if it wishes, exclude those with differing views. The result will be two or more independent ethical frameworks and infinite potential for moral posturing during conflicts. One need not look far to see this theory borne out by current events.

18. Following Barthes, I use myth within the Saussurean signification system wherein the union of the signifier and signified yields a sign, imbued with meaning through the communicative act. Barthes uses the example of a black pebble as potential signifier. When used as the indicator of a death sentence (the signified), the black pebble, as the union of signifier and signified, becomes a sign. But the sign is only valid within the specific context (myth) that gives it meaning (*Mythologies* 111). And myth, as Barthes repeatedly stresses, is a "type of speech," or language (109).

19. See chapter 6 for an extended discussion of sustainability as I envision it defining a post-Reclamation reality. While I do not disagree with Luke, my definition of sustainability requires multiple perspectives and would not permit one dominant interest to transform it into an "economic project."

20. Both the dam and the cycle form signature features of faux potlatch. Systems theory presupposes a form of potlatch that recalls Bataille: "All structures of social systems have to be based on this fundamental fact of vanishing events . . . or words that are dying away. . . . The events themselves cannot be saved; their loss is the condition of their regeneration. . . . Disintegration and reintegration, disordering and reordering require each other, and reproduction comes about only by a recurring integration of disintegration and reintegration" (Luhmann, *Essays* 9). Since faux potlatch displaces the normal acquisition/expenditure cycle, it is not surprising to find it at work in a system dedicated to limitless growth.

21. Systems are not capable of long-range planning. Hence they are always concerned with short-term autopoiesis. Other matters become relevant once survival is assured. "[O]nce the essential ontogenic requisite of reproduction is fulfilled, everything is made possible. Reproduction is a necessity; otherwise there is extinction" (Maturana and Varela, *Tree* 107). Capital is necessary to the system's short-term survival, whereas the system can sustain a great deal of ecological damage and still endure. Therefore, capital acquisition takes precedence.

22. Big 10 members include Environmental Defense Fund, Environmental Policy Institute, Friends of the Earth, Izaak Walton League of America, National Audubon Society, National Parks and Conservation Association, National Wildlife Federation, Natural Resources Defense Council, Sierra Club, and the Wilderness Society.

23. In an early essay applying systems theory to ecology and the notion of environment, Bernard Patten recognizes and rejects the notion of prediction in system-environment interaction: "[T]he [system] cannot anticipate its future environment; it is *nonanticipatory*. . . . [T]he only way it could [predict] (since it is determinate) would be based on information about the future. This possibility is precluded for the causal object" (259).

24. Observing environmental interaction requires an ongoing calculus that strives to integrate entropic variations. Yet the calculus is an elaborate fiction; only one outcome can result from a given perturbation—the one that actually occurs. As Luhmann observes, "The theory of self-referential systems . . . has realized that the classical instruments of the acquisition of knowledge, namely deduction (logic) and causality (experience), are merely forms of simplifying the observation of observations. . . . Methodologically, this means that the point of departure has to be the observation of self-observing systems and not the assumed logic of causal-

ity" (*Communication* 9). Calculating probabilities is a second-order organizational device (i.e., derived from a system's ability to observe itself) and an implicit admission that though only one possibility exists, we cannot know it.

25. Maturana and Varela do distinguish, however, between language as acquired behavior and the innate behavior seen in the insects. They note that acquired communicative behaviors depend on an organism's ontogeny as well as its particular history of social interactions. Innate behavior, by contrast, can characterize an entire class of organisms and is independent of ontogeny and social history (*Tree* 207).

26. The principle of the actor as observer is integral to second-order cybernetics and essential to the application of systems-theoretical principles to human social systems. See Rasch and Wolfe, especially 7–13.

27. As Luhmann succinctly notes, "A system can see only what it can see. It cannot see what it cannot. Moreover, it cannot see that it cannot see this" (*Communication* 23). Interestingly, defining the unknown simultaneously demarcates the known. It therefore constitutes an act of self-definition (*Essays* 7).

28. See, for example, Loyd's decision to give up cockfighting after Codi persuades him that animals suffering for human amusement is incongruent with the dignity of life (*Animal* 185–91).

29. Maturana and Varela see language as the fundamental constitutive force defining the human condition. As such, it is virtually all-powerful: "Every human act takes place in language. Every act in language brings forth a world created with others in the act of coexistence which gives rise to what is human. Thus every human act has an ethical meaning because it is an act of constitution of the human world" (*Tree* 247).

30. It is an often-remarked-upon phenomenon that the most powerful art tends to come from society's margins. From a systems-theoretical perspective, that trend is entirely logical; art objects are communicative devices and communication arises in response to disturbance. Disturbance will not likely come from components of a function system thriving within the current equilibrium. It will rather be generated by those segments of society whose needs and goals conflict in some capacity with those of the unity. Art upsets equilibrium through interrogating the known and offering new ways of seeing. In other words, it disturbs various societal function systems, generating resonance and forcing them to adapt and evolve. *Animal Dreams* is an example of one such self-conscious attempt to force beneficial adaptation in the social system currently governing the West.

31. As Evernden notes, "Ecology cannot be presumed to be the exclusive ally of the environmental movement, for it provides information that can just as well be used to manipulate nature as to defend it" (10).

Chapter 6

1. Nicholas Wade comments, "Given that consciousness, the one thing that we feel absolutely sure about, is in fact an illusion, it is not surprising that we lack the vocabulary to describe this embarrassing state of affairs" (22).

2. This phenomenon extends to the legal realm as well. Robin West, among others, notes that the legal system's control over the means of normative discourse creates "clusters of beliefs" in the legitimacy of the social structures that buttress the society:

> One part of those clusters of beliefs is the moral perspective . . . from which particular laws are criticized or celebrated. For example, while we may feel free to criticize a particular legal rule as inefficient, or a proposed bill as nonrepresentative of the public will, or a criminal statute as overintrusive into our privacy, we find it much harder to criticize our moral commitment to efficiency, or representation, or privacy against which these laws are evaluated. . . . The result is that the vast bulk of the particular rules and the process of the extant system that govern our behavior are seen as morally legitimate—as in accord with our moral beliefs. Meaningful criticism of the law against truly independent moral standards is thereby frustrated. (5–6)

As with environmentalism, critics of the legal system have consequently been frustrated in their attempts to articulate "a credible basis for the moral criticism of law that would be other than complicit in the legitimation of the law's fundamental authority" (6).

3. Latour defines the discursive constitution as the set of practical and theoretical principles that "defines humans and nonhumans, their properties and their relations, their abilities and their groupings" (15).

4. See for example, Porush; Landow; and Paulsen.

5. Speaking of the inevitable conflicts of multiculturalism, Stanley Fish notes that substantial numbers of citizens of one country are finding themselves also members of a culture that calls into question that country's "philosophical boundaries." This collision of values means that there may never exist any permanent formula for resolving conflicts between and among citizens. What is needed, Fish argues, citing Charles Taylor, is some "inspired adhoccery" (63). This notion of "inspired adhoccery" aptly captures the essence of Sustainability as I envision it.

6. There is some debate among theorists as to whether hypertext can properly be considered nonlinear. It is certainly true that many if not most hypertext users often follow highly linear paths, screening out information that is not already identified and defined. Martin Rosenberg argues that

hypertext is, for this reason, better labeled "multilinear." While Rosenberg is correct that hypertext, at least for the moment, does retain significant vestiges of linearity, it clearly offers an exponentially larger number of narrative possibilities.

7. No one has summed up our ecological dilemma more concisely than the framers of the Endangered Species Act. The 1973 House report accompanying the bill declared, "Man's . . . effective domination over the world's support systems has taken place within a few short generations. Our ability to destroy, or almost destroy, all intelligent life on the planet became apparent only in this generation. A certain humility, and a sense of urgency, seem indicated" (United States Congress).

8. The map analogy is not wholly accurate because the system and environment are not finite. Even as the system grows more complex and gains a little territory from the environment, so too does the environment grow more complex and regain its previous size. A more accurate analogy might be a three-dimensional map with the system and environment sharing one border but having nothing limiting their expansion on any other side.

9. Donna Haraway's "Cyborg Manifesto" provides a fascinating and provocative discussion of the boundaries between humans and technology. See also Jim Tarter, "Collective Subjectivity and Postmodern Ecology," for an interesting analysis of the impact of fluid boundaries on the discipline of ecology.

10. Wallace Stegner describes humanity's dilemma beautifully: "We are a wild species, as Darwin pointed out. Nobody ever tamed or domesticated or scientifically bred us. But for at least three millennia we have been engaged in a cumulative and ambitious race to modify and gain control of our environment, and in the process we have come close to domesticating ourselves. . . . Just as surely as it has brought us increased comfort and more material goods, [progress] has brought us spiritual losses, and now threatens to become the Frankenstein that will destroy us" (*Sound* 145).

11. Chenoweth insists upon being addressed as "Congress*man*" because she believes it is the only form of the title that commands respect. She was reelected in 1996 and 1998.

12. The mechanistic paradigm was developed in the 1620s and 1630s by a group of French thinkers including Descartes, Mersenne, and Gassendi. They drew on the Baconian principle of dominion over nature to redefine contemporary notions of order. Instead of picturing the world as an organic whole, functioning according to the will of nature, they pictured it mechanically, wherein each part functioned according to a rational and determinable set of laws. Once the machine replaced the infinite cosmos as the metaphor for nature, it became possible to imagine tinkering with the mechanism and reshaping the world according to principles of human rationality. For a fas-

cinating discussion of the development and subsequent dominance of the mechanistic paradigm, see Merchant, *Death* 192–235.

13. There is considerable disagreement among scholars as to the democratizing properties of hypermedia. See Charles Ess, "The Political Computer," especially pages 230–32, for a survey of the differing views. For my purposes, the debate is not relevant. Whether hypertext truly democratizes or merely creates a new hierarchy, Sustainability's efficacy remains unimpaired.

14. According to *Wired* magazine, even corporate CEOs have signed on. The October 1996 issue features an interview with author and former Citicorp/Citibank CEO Walter Wriston. While Wriston "used to be the most powerful banker in the world," according to the interviewer, "now he's talking like a cyberpunk" (Bass). These types of linked linguistic and social shifts lend credence to Richard Rorty's contention that cultural change is more about changing vocabularies than unearthing truths (*Philosophy* 352).

15. This development proves especially important because both ends of the ideological spectrum are prone to appropriating the mantle of objectivity. Mechanistic science has long assumed itself to be ideology-free, while feminist philosophers of science like Sandra Harding oppose traditional models by arguing for a "strong objectivity" capable of culling disruptive (anti) social interests from research data. See Wolfe 40–41.

16. Deep ecologists, for example, are often accused of misanthropy because one of the movement's principal tenets is that of biocentric equality—that "all things in the biosphere have an equal right to live and blossom and . . . that all organisms and entities in the ecosphere are . . . equal in intrinsic worth" (Devall and Sessions 67). Allegiance to this view can lead to troublesome ontological questions about the relative value of the malaria mosquito or the living virus that causes AIDS, compared to the value of human life.

The deep ecological perspective has come under withering criticism even from within the environmental movement. See, for example, Murray Bookchin's critique in *Remaking Society*, 9–13.

17. Among the linked causes of environmentalism that might wind up on opposite sides of the incinerator issue are groups that support incineration over burying wastes and their counterparts who support ground storage; indigenous rights groups versus advocates of environmental justice; wilderness preservationists (if the alternative to incineration involves storing the waste in a remote area) versus proponents of urban renewal; and so on. The incinerator, like many issues that theoretically should galvanize the environmental movement, might polarize it instead.

18. Walter Wriston argues that the basis for wealth has evolved from land to labor to information. In a famous remark now inscribed in the lobby of the

lobby of New York's Library of Science, Industry, and Business, Wriston declared, "Information about money has become almost as important as money itself." Since information cannot actually be sold, this thought might more accurately be expressed as "*Access to* information about money has become almost as important as money itself."

19. Fredric Jameson addresses the dilemma of capitalism in an information society, claiming that the distillation of living entities into bodies of information actually increases commodification rather than complicating or diminishing it.

20. According to Jack Stanford, director of the University of Montana's Flathead Lake Biological Station, "The river is essentially dead. You can put a drift net in the river for fifteen minutes and catch one or two insects" (qtd. in Devine 72–73).

21. Sam Howe Verhovek, "Returning River to Salmon and Man to the Drawing Board," *New York Times,* September 25, 1999, 28.

22. There have been other noteworthy developments in the realm of dam demolition. In the summer of 1999, a hydroelectric dam on Maine's Kennebec River was demolished, per federal order, in order to open the river to salmon and sturgeon migration. In addition, Pacificorp, a northwestern utility company, recently announced that it would take down the Condit Dam on Washington's White Salmon River by the year 2006 rather than pay $30 million to retrofit it to enable salmon migration. The Condit, at 125 feet high, is the tallest dam ever slated for demolition in the United States (Verhovek, 28).

23. Timothy Luke offers an analogous blueprint for reform that he calls "ecological populism." Luke's formulation, grounded in a critique of the ecopolitics of Bookchin, Marcuse, and Soleri, offers a vision of "communities . . . united by new narratives of their own historical consciousness, beyond and behind the nationalist myths of new class bureaucrats nor the progress discourses of corporate public relations, and by more concretely articulated social, ecological, and cultural interests grounded in immediate environmental and political conditions" (*Ecocritique* 200).

24. I adapt these ideas from a talk Mitchell delivered at the 1995 Digital Dialectic Conference in Pasadena, California. See also *City of Bits,* a book that appears in traditional print form from MIT Press, or in hypertext at http://www.mitpress.mit.edu/City_of_Bits/index.html.

Bibliography

Abbey, Edward. *Abbey's Road.* New York: Penguin, 1991.

———. *Beyond the Wall.* New York: Holt, Rinehart and Winston, 1984.

———. *Black Sun.* New York: Simon and Schuster, 1971.

———. *The Brave Cowboy.* New York: Avon, 1992.

———. *Desert Solitaire: A Season in the Wilderness.* New York: Ballantine, 1978.

———. *Down the River.* New York: E. P. Dutton, 1982.

———. *Fire on the Mountain.* New York: Avon, 1992.

———. *The Fool's Progress: An Honest Novel.* New York: Avon, 1990.

———. *Good News.* New York: Plume, 1991.

———. *Hayduke Lives!* Boston: Little, Brown, 1990.

———. *The Journey Home.* New York: E. P. Dutton, 1977.

———. *The Monkey Wrench Gang.* New York: Avon, 1976.

———. *One Life at a Time, Please.* New York: Henry Holt, 1988.

Abram, David. *The Spell of the Sensuous.* New York: Vintage, 1997.

Adams, Henry. *The Education of Henry Adams.* New York: Random House, 1931.

Allen, John L. "The Garden-Desert Continuum: Competing Views of the Great Plains in the Nineteenth Century." *Great Plains Quarterly* 5.4 (fall 1985): 207–20.

Ambrose, Stephen. *Undaunted Courage: Meriwether Lewis, Thomas Jefferson, and the Opening of the American West.* New York: Simon and Schuster, 1996.

Austin, Mary. *Cactus Thorn: A Novella.* Reno: University of Nevada Press, 1988.

———. *Earth Horizon.* Boston: Houghton Mifflin, 1932.

———. *Stories from the Country of Lost Borders.* New Brunswick: Rutgers University Press, 1987.

———. *The Ford.* Berkeley and Los Angeles: University of California Press, 1997.

———. *The Land of Journey's Ending.* New York: Century, 1924.

———. *The Land of Little Rain.* New York: Penguin, 1988.

Bacon, Francis. *Essays and New Atlantis.* London: Walter J. Black, 1942.

Baltensperger, Bradley H. *Nebraska: A Geography.* Boulder, Colo.: Westview Press, 1985.

Barthes, Roland. *A Barthes Reader.* Ed. Susan Sontag. New York: Hill and Wang, 1983.

———. *Mythologies.* Trans. Jonathan Cape. New York: Hill and Wang, 1982.

Bass, Thomas A. "The Future of Money." *Wired* 4.10 (October 1996). www.wired.com/wired/archive/4.10.

Bataille, Georges. *Visions of Excess: Selected Writings, 1927–1939.* Ed. and trans. Allan Stoekl with Carl R. Lovitt and Donald M. Lesle Jr. Minneapolis: University of Minnesota Press, 1985.

Bateson, Gregory. *Steps to an Ecology of Mind.* New York: Ballantine, 1990.

Baudrillard, Jean. *America.* Trans. Chris Turner. New York: Verso, 1988.

———. *The Ecstasy of Communication.* Ed. Slyveree Lotringer, trans. Bernard Schutze and Caroline Schutze. New York: Autonomedia, 1988.

———. *The Mirror of Production.* Trans. Mark Poster. St. Louis: Telos Press, 1975.

———. *Selected Writings.* Ed. and trans. Mark Poster. Stanford: Stanford University Press, 1988.

Beegel, Susan F., Susan Shillinglaw, and Wesley N. Tiffney Jr., eds. *Steinbeck and the Environment: Interdisciplinary Approaches.* Tuscaloosa: University of Alabama Press, 1997.

Belfrage, Cedric. *The Promised Land.* New York: Garland, 1978.

Benson, Jackson. "Hemingway the Hunter and Steinbeck the Farmer." *Michigan Quarterly Review* 24.3 (summer 1985): 441–60.

———. "Through a Political Glass, Darkly: The Example of John Steinbeck." *Studies in American Fiction* 12.1 (spring 1984): 45–59.

———. *The True Adventures of John Steinbeck, Writer.* New York: Viking, 1984.

Benton, Thomas Hart. "Discourse of Mr. Benton of Missouri before the Boston Mercantile Library Association on the Physical Geography of the Country between the States of Missouri and California." Tremont Temple, Boston, December 20, 1854.

Berkman, Richard L., and W. Kip Viscusi. *Damming the West.* New York: Grossman Press, 1973.

Berry, Wendell. *What Are People For?* San Francisco: North Point, 1990.

Bertalanffy, Ludwig von. *Perspectives on General System Theory.* Ed. Edgar Taschdjian. New York: George Braziller, 1975.

Billington, Ray Allen. *Westward Expansion: A History of the American Frontier.* 4th ed. New York: Macmillan, 1974.

Bishop, James, Jr. *Epitaph for a Desert Anarchist: The Life and Legacy of Edward Abbey*. New York: Atheneum, 1994.

Blumenthal, Sidney. "Her Own Private Idaho." *New Yorker* July 10, 1995: 27–33.

Bookchin, Murray. *Remaking Society: Pathways to a Green Future*. Boston: South End Press, 1990.

Bookchin, Murray, and Dave Foreman. *Defending the Earth: A Dialogue between Murray Bookchin and Dave Foreman*. Boston: South End Press, 1991.

Bramwell, Anna. *Ecology and the Twentieth Century: A History*. New Haven: Yale University Press, 1989.

Bryant, Paul L. "Edward Abbey and Environmental Quixoticism." *Western American Literature* 24.1 (May 1989): 37–43.

Callicott, J. Baird. *In Defense of the Land Ethic: Essays in Environmental Philosophy*. Albany: State University of New York Press, 1989.

Carson, Rachel. *Silent Spring*. Boston: Houghton Mifflin, 1987.

Cassuto, David N. *Cold Running River*. Ann Arbor: University of Michigan Press, 1994.

Cassuto, Leonard. *The Inhuman Race*. New York: Columbia University Press, 1997.

Cole, Douglas, and Ira Chaikin. *An Iron Hand upon the People: The Law against Potlatch on the Northwest Coast*. Seattle: University of Washington Press, 1990.

Collard, Andrée, and Joyce Contrucci. *Rape of the Wild: Man's Violence against Animals and the Earth*. Bloomington: Indiana University Press, 1989.

Crawford, John F., and Annie O. Eysturoy, eds. *This Is about Vision: Interviews with Southwestern Writers*. Albuquerque: University of New Mexico Press, 1990.

Crawford, Stanley. *Mayordomo: Chronicle of an Acequia in Northern New Mexico*. Albuquerque: University of New Mexico Press, 1993.

Cronon, William, ed. *Uncommon Ground: Towards Reinventing Nature*. New York: Norton, 1995.

Darwin, Charles. *The Descent of Man*. Princeton: Princeton University Press, 1981.

———. *On the Origin of Species*. Cambridge: Harvard University Press, 1964.

Davis, Robert Con, ed. *The Grapes of Wrath: A Collection of Critical Essays*. Englewood Cliffs, N.J.: Prentice-Hall, 1982.

Devall, Bill, and George Sessions, eds. *Deep Ecology: Living As If Nature Mattered*. Salt Lake City: Gibbs Smith, 1985.

Devine, Robert S. "The Trouble with Dams." *Atlantic Monthly* (August 1995): 64–74.

Ditsky, John, ed. *Critical Essays on Steinbeck's "The Grapes of Wrath."* Boston: Hall, 1989.

Douglas, Mary. *Purity and Danger.* London: Routledge and Kegan Paul, 1966.

Eagleton, Terry. *The Ideology of the Aesthetic.* Cambridge, Mass.: Basil Blackwell, 1990.

Eco, Umberto. *Semiotics and the Philosophy of Language.* Bloomington: Indiana University Press, 1986.

Emerson, Ralph Waldo. *Essays and Lectures.* New York: Library of America, 1983.

———. *Selections from Ralph Waldo Emerson.* Ed. Stephen E. Whicher. Boston: Houghton Mifflin, 1957.

———. *The Works of Ralph Waldo Emerson.* Vol. 1. Boston: Houghton and Mifflin, 1876.

Emmons, David M. *Garden in the Grasslands: Boomer Literature of the Central Great Plains.* Lincoln: University of Nebraska Press, 1972.

Ess, Charles. "The Political Computer." *Hypertext Theory.* Ed. George P. Landow. Baltimore: Johns Hopkins University Press, 1994.

Evernden, Neil. *The Social Creation of Nature.* Baltimore: Johns Hopkins University Press, 1992.

Fink, Augusta. *I-Mary: A Biography of Mary Austin.* Tucson: University of Arizona Press, 1983.

Fish, Stanley. *The Trouble with Principle.* Cambridge: Harvard University Press, 1999.

Fitzgerald, F. Scott. *The Great Gatsby.* New York: Scribner and Sons, 1925.

Flader, Susan. *Thinking Like a Mountain.* Columbia: University of Missouri Press, 1984.

Foreman, Dave. *Confessions of an Eco-Warrior.* New York: Harmony, 1991.

Foreman, Dave, and Bill Haywood, eds. *Ecodefense: A Field Guide to Monkey Wrenching.* Tucson: N. Ludd Press, 1988.

Foucault, Michel. *The Order of Things: An Archeology of the Human Sciences.* Trans. Alan Sherdan-Smith. New York: Vintage, 1970.

———. *Power/Knowledge: Selected Interviews and Other Writings, 1972–1977.* Ed. Colin Gordon. New York: Pantheon, 1980.

Fradkin, Philip L. *A River No More.* New York: Knopf, 1981.

Fritzell, Peter A. *Nature Writing and America.* Ames: Iowa State University Press, 1990.

Habermas, Jürgen. *Legitimation Crisis.* Trans. Thomas McCarthy. Boston: Beacon Press, 1975.

Getches, David. *Water Law.* 3d ed. St. Paul: West, 1997.

Gilpin, William. *The Cosmopolitan Railway, Compacting and Fusing Together All the World's Continents.* San Francisco: History Co., 1890.

———. *The Grain, Pastoral, and Gold Regions of North America.* Philadelphia: Sower Barnes and Co., 1860.

———. *The Mission of the North American People: Geographical, Social, and Political.* 2d ed. Philadelphia: J. B. Lippincott, 1874.

Glotfelty, Cheryll, and Harold Fromm, eds. *The Ecocriticism Reader.* Athens: University of Georgia Press, 1996.

Gore, Al. *Earth in the Balance: Ecology and the Human Spirit.* New York: Houghton Mifflin, 1992.

Gorz, Andre. *Ecology as Politics.* Trans. Patsy Vigderman and Jonathan Cloud. Boston: South End Press, 1980.

Gould, Stephen Jay. *The Mismeasure of Man.* New York: Norton, 1981.

Gragg, Frances, and William Putnam. *The Golden Valley: A Novel of California.* New York: Duell, Sloan and Peoree, 1950.

Gregory, James N. *American Exodus: The Dust Bowl Migration and Okie Culture in California.* New York: Oxford University Press, 1989.

Greiner, Patricia. "Radical Environmentalism in Recent Literature concerning the American West." *Rendezvous* 19.1 (fall 1983): 8–15.

Griffin, Susan. *Woman and Nature: The Roaring Inside Her.* New York: Harper, 1978.

Gross, Alan. *The Rhetoric of Science.* Cambridge: Harvard University Press, 1990.

Grusin, Richard. "Thoreau, Extravagance, and the Economy of Nature." *American Literary History* 5.1 (spring 1993): 30–50.

Haney-López, Ian F. *White by Law: The Legal Construction of Race.* New York: New York University Press, 1996.

Haraway, Donna J. "Cyborg Manifesto." *Simians, Cyborgs, and Women: The Reinvention of Nature.* New York: Routledge, 1991.

Hardin, Garrett, assembler. *Population, Evolution, and Birth Control: A Collage of Controversial Ideas.* 2d ed. San Francisco: W. H. Freeman, 1969.

Harvey, David. *The Condition of Postmodernity: An Inquiry into the Origins of Cultural Change.* Cambridge: Blackwell, 1989.

Hayles, Katherine. "Searching for Common Ground." *Reinventing Nature: Responses to Postmodern Deconstruction.* Ed. Michael E. Soulé and Gary Lease. Washington, D.C.: Island, 1995.

———. "Simulated Nature and Natural Simulations: Rethinking the Relation between the Beholder and the Natural World." *Uncommon Ground: Towards Reinventing Nature.* Ed. William Cronon. New York: Norton, 1995.

———. "Virtual Bodies and Flickering Signifiers." *October* 66 (fall 1993): 69–91.

Hepworth, James, and Gregory McNamee, eds. *Resist Much, Obey Little: Some Notes on Edward Abbey.* Salt Lake City: Dream Garden, 1989.

Horgan, Paul. *Of America East and West.* New York: Farrar, Straus and Giroux, 1984.

———. "The Western Novel—a Symposium." *South Dakota Review* autumn 1964.

Hofstadter, Richard. *The Age of Reform from Bryan to FDR.* New York: Knopf, 1972.

Hoyt, John. *Drought of 1930–34.* U.S. Geological Survey, Water Supply Paper 680. Washington, D.C., 1936.

Hundley, Norris, Jr. *The Great Thirst: Californians and Water, 1770s–1990s.* Berkeley and Los Angeles: University of California Press, 1992.

———. *Water and the West: The Colorado River Compact and the Politics of Water in the American West.* Berkeley and Los Angeles: University of California Press, 1975.

Illich, Ivan. *H_2O and the Water of Forgetfulness.* Berkeley, Calif.: Heyday Press, 1985.

Jameson, Fredric. *The Political Unconscious: Narrative as a Socially Symbolic Act.* Ithaca, N.Y.: Cornell University Press, 1981.

———. *Postmodernism; or, The Cultural Logic of Late Capitalism.* Durham: Duke University Press, 1991.

Jay, Gregory. *America the Scrivener: Deconstruction and the Subject of Literary History.* Ithaca, N.Y.: Cornell University Press, 1990.

Jehlen, Myra. *American Incarnation: The Individual, the Nation, and the Continent.* Cambridge: Harvard University Press, 1986.

Kerr, Robert S. *Land, Wood, and Water.* New York: Fleet, 1960.

Kingsolver, Barbara. *Animal Dreams.* New York: Harper Perennial, 1991.

———. *The Bean Trees.* New York: Harper and Row, 1988.

———. *High Tide in Tucson: Essays from Now or Never.* New York: HarperCollins, 1995.

———. *Holding the Line: Women in the Great Arizona Mine Strike of 1983.* Ithaca, N.Y.: ILR Press, 1989.

———. *Homeland and Other Stories.* New York: Harper and Row, 1989.

———. *Pigs in Heaven.* New York: HarperCollins, 1983.

Kolodny, Annette. *The Land before Her: Fantasy and Experience of the American Frontiers, 1630–1860.* Chapel Hill: University of North Carolina Press, 1984.

———. *The Lay of the Land.* Chapel Hill: University of North Carolina Press, 1975.

Krutch, Joseph Wood. *The Desert Year.* New York: William Sloane, 1952.
Kupperman, Karen Ordahl. "Fear of Hot Climates in the Anglo-American Colonial Experience." *William and Mary Quarterly* 41.2 (1984): 213–40.
———. "The Puzzle of the American Climate in the Early Historical Period." *American Historical Review* 87.5 (December 1982): 1262–89.
La Porte, Todd R. "Organized Social Complexity: Explication of a Concept." *Organized Social Complexity: Challenge to Politics and Policy.* Ed. Todd La Porte. Princeton: Princeton University Press, 1975.
Landow, George P. *Hypertext: The Convergence of Contemporary Critical Theory and Technology.* Baltimore: Johns Hopkins University Press, 1992.
Lange, Dorothea, and Paul Schuster Taylor. *An American Exodus: A Record of Human Erosion in the Thirties.* New Haven: Yale University Press, 1969.
Lanham, Richard A. *The Electronic Word: Democracy, Technology, and the Arts.* Chicago: University of Chicago Press, 1993.
Latour, Bruno. *We Have Never Been Modern.* London: Harvester Wheatsheaf, 1993.
Leopold, Aldo. *A Sand County Almanac.* New York: Ballantine, 1988.
Lerner, Gerda. *The Creation of Feminist Consciousness.* New York: Oxford University Press, 1993.
Levine, George, ed. *One Culture: Essays in Science and Literature.* Madison: University of Wisconsin Press, 1987.
Lewis, Martin W. *Green Delusions: An Environmentalist Critique of Radical Environmentalism.* Durham: Duke University Press, 1992.
Lewis, Meriwether, and William Clark. *The Journals of the Lewis and Clark Expedition.* Lincoln: University of Nebraska Press, 1983.
Light, Andrew. "Urban Wilderness." *Wild Ideas.* Ed. David Rothenberg. Minneapolis: University of Minnesota Press, 1995.
Lopez, Barry. *Crossing Open Ground.* New York: Scribner, 1988.
Luhmann, Niklas. *Ecological Communication.* Trans. John Bednarz Jr. Chicago: University of Chicago Press, 1986.
———. *Essays on Self-Reference.* New York: Columbia University Press, 1990.
———. *A Sociological Theory of Law.* Trans. Elizabeth King and Martin Albrow. Boston: Routledge and Kegan Paul, 1985.
Luke, Timothy W. *Ecocritique: Contesting the Politics of Nature, Economy, and Culture.* Minneapolis: University of Minnesota Press, 1997.
———. "On Environmentality: Geo-power and Eco-knowledge in the Discourses of Contemporary Environmentalism." *Cultural Critique* 30 (spring 1995): 57–82.

Lyotard, Jean-François. *The Inhuman: Reflections on Time.* Trans. Geoffrey Bennington and Rachel Bowlby. Stanford: Stanford University Press, 1991.

Manes, Christopher. *Green Rage: Radical Environmentalism and the Unmaking of Civilization.* Boston: Little, Brown, 1990.

Marsh, George Perkins. *Man and Nature; or, Physical Geography as Modified by Human Action.* Cambridge: Harvard University Press, 1965.

Marx, Karl. *The Economic and Philosophic Manuscripts of 1844.* Ed. Dirk J. Struik, trans. Martin Milligan. New York: International, 1964.

Marx, Leo. *The Machine in the Garden.* New York: Oxford University Press, 1964.

———. "Pastoralism in America." *Ideology and Classic American Literature.* Ed. Sacvan Bercovitch and Myra Jehlen. New York: Cambridge University Press, 1986.

Maturana, Humberto R. "Science and Daily Life: The Ontology of Scientific Explanation." *Research and Reflexivity.* Ed. Frederick Steier. London: Sage, 1991.

Maturana, Humberto R., and Francisco J. Varela. *Autopoeisis and Cognition: The Realization of the Living.* Dordrecht: D. Reidel, 1985.

———. *The Tree of Knowledge: The Biological Roots of Human Understanding.* Trans. Robert Paolucci. Boston: Shambhala, 1992.

McKibben, Bill. *The End of Nature.* New York: Random House, 1989.

McClintock, James I. "Edward Abbey's 'Antidotes to Despair.'" *Studies in Contemporary Fiction* 31.1 (fall 1989): 41–54.

McPhee, John. *The Control of Nature.* New York: Farrar, Straus and Giroux, 1989.

———. *Encounters with the Archdruid.* New York: Farrar, Straus and Giroux, 1971.

McWilliams, Carey. *Factories in the Field: The Story of Migratory Farm Labor in California.* Santa Barbara: Peregrine Press, 1971.

Merchant, Carolyn. *The Death of Nature: Women, Ecology, and the Scientific Revolution.* San Francisco: Harper and Row, 1983.

———. *Ecological Revolutions: Nature, Gender, and Science in New England.* Chapel Hill: University of North Carolina Press, 1989.

———. *Radical Ecology: The Search for a Livable World.* New York: Routledge, 1992.

Mitchell, Don. *The Lie of the Land: Migrant Workers and the California Landscape.* Minneapolis: University of Minnesota Press, 1996.

Muir, John. *My First Summer in the Sierra.* New York: Penguin, 1987.

———. *The Yosemite.* Garden City: Doubleday, 1962.

Murray, John A. "The Hill beyond the City: Elements of the Jeremiad in Edward Abbey's 'Down the River with Henry Thoreau.'" *Western American Literature* 22.4 (February 1988): 301–6.

Nash, Roderick. *The Rights of Nature: A History of Environmental Ethics.* Madison: University of Wisconsin Press, 1989.

———. *Wilderness and the American Mind.* 3d ed. New Haven: Yale University Press, 1989.

———, ed. *American Environmentalism: Readings in Conservation History.* 3d ed. New York: McGraw-Hill, 1990.

National Park Foundation. *Mirror of America: Literary Encounters with National Parks.* Boulder, Colo.: Roberts Reinhart, 1989.

National Resource Council. *Restoration of Aquatic Ecosystems.* Washington, D.C.: National Academy Press, 1992.

Nichols, John. *The Magic Journey.* New York: Ballantine, 1978.

———. *The Milagro Beanfield War.* New York: Ballantine, 1974.

———. *The Nirvana Blues.* New York: Ballantine, 1981.

Norwood, Vera. *Made from This Earth: American Women and Nature.* Chapel Hill: University of North Carolina Press, 1993.

Owens, Louis. *John Steinbeck's Re-vision of America.* Athens: University of Georgia Press, 1985.

Patten, Bernard C. "Systems Approach to the Concept of Environment." *General Systems: Yearbook of the Society for General Systems Research.* Vol. 24. Ed. Brian Gaines. Louisville: Society for General Systems Research, 1979.

Paulsen, William R. *The Noise of Culture: Literary Texts in a World of Information.* Ithaca, N.Y.: Cornell University Press, 1988.

Pierce, Christine, and Donald VanDeVeer, eds. *People, Penguins, and Plastic Trees: Basic Issues in Environmental Ethics.* Belmont: Wadsworth, 1986.

Porush, David. *The Soft Machine: Cybernetic Fiction.* New York: Methuen, 1985.

Postman, Neil. *Technopoly: The Surrender of Culture to Technology.* New York: Vintage, 1992.

Potter, Loren D., and Charles L. Drake. *Lake Powell: Virgin Flow to Dynamo.* Albuquerque: University of New Mexico Press, 1989.

Powell, John Wesley. *The Exploration of the Colorado River and Its Canyons.* New York: Dover, 1961.

———. "Institutions for the Arid Lands." *Century* 39 (May 1890): 111–16.

———. "Irrigable Lands of the Arid Region." *Century* 39 (March 1890): 766–76.

———. "Non-irrigable Lands of the Arid Region." *Century* 39 (April 1890): 915–22.

———. *Report on the Lands of the Arid Region of the United States.* 2d ed. Washington, D.C.: Government Printing Office, 1879.

———. *Selected Prose of John Wesley Powell.* Ed. George Crossette. Boston: David R. Godine, 1970.

Press, Daniel. *Democratic Dilemmas in the Age of Ecology: Trees and Toxics in the American West.* Durham: Duke University Press, 1994.

Rasch, William, and Cary Wolfe. "Theory of a Different Order: A Conversation with Katherine Hayles and Niklas Luhmann." *Cultural Critique* 30 (spring 1995): 7–36.

Rawls, John. *The Law of Peoples.* Cambridge: Harvard University Press, 1999.

Reisner, Marc. *Cadillac Desert: The American West and Its Disappearing Water.* Rev. ed. New York: Penguin, 1993.

Roach, Catherine. "Loving Your Mother: On the Woman-Nature Relation." *Hypatia* 6.2 (1991): 46–59.

Rogin, Michael Paul. "Nature as Politics and Nature as Romance in America." *Political Theory* 5.1 (February 1977): 5–30.

Ronald, Ann. *The New West of Edward Abbey.* Albuquerque: University of New Mexico Press, 1982.

Rorty, Richard. *Objectivity, Relativism, and Truth.* Cambridge: Cambridge University Press, 1991.

———. *Philosophy and the Mirror of Nature.* Princeton: Princeton University Press, 1979.

Rosenberg, Martin E. "Physics and Hypertext: Liberation and Complicity in Art and Pedagogy." *Hypertext Theory.* Ed. George P. Landow. Baltimore: Johns Hopkins University Press, 1994.

Rosman, Abraham, and Paula G. Rubel. *Feasting with Mine Enemy: Rank and Exchange among Northwest Coast Societies.* New York: Columbia University Press, 1971.

Ross, Andrew. *Strange Weather: Culture, Science, and Technology in the Age of Limits.* London: Verso, 1991.

Rothenberg, David, ed. *Wild Ideas.* Minneapolis: University of Minnesota Press, 1995.

Rousseau, Jean-Jacques. *Discourse on Political Economy and the Social Contract.* Trans. Christopher Betts. New York: Oxford University Press, 1994.

Rubin, Alfred. "Pollution by Analogy: The Trail Smelter Arbitration." *Oregon Law Review* 50 (1971): 259–82.

Rudnick, Lois. "Re-naming the Land: Anglo-Expatriate Women in the Southwest." *The Desert Is No Lady: Southwestern Landscapes in*

Women's Writing and Art. Ed. Vera Norwood and Janice Monk. New Haven: Yale University Press, 1987.

Sax, Joseph L., Robert H. Abrams, and Barton H. Thompson Jr. *Legal Control of Water Resources: Cases and Materials*. 2d. ed. St. Paul: West, 1991.

Sears, John F. *Sacred Places: American Tourist Attractions in the Nineteenth Century*. New York: Oxford University Press, 1989.

Seltzer, Mark. *Bodies and Machines*. New York: Routledge, 1992.

Slotkin, Richard. *The Fatal Environment: The Myth of the Frontier in the Age of Industrialization, 1800–1890*. New York: Atheneum, 1985.

———. *Regeneration through Violence: The Mythology of the American Frontier, 1600–1860*. Middletown, Conn.: Wesleyan University Press, 1973.

Smith, Henry Nash. *Virgin Land: The American West as Symbol and Myth*. New York: Vintage Books, 1950.

Smith, Zachary, ed. *Water and the Future of the Southwest*. Albuquerque: University of New Mexico Press, 1989.

Stegner, Wallace. *The American West as Living Space*. Ann Arbor: University of Michigan Press, 1987.

———. *Beyond the Hundredth Meridian: John Wesley Powell and the Second Opening of the American West*. Lincoln: University of Nebraska Press, 1982.

———. *The Sound of Mountain Water: The Changing American West*. New York: E. P. Dutton, 1980.

———. *Where the Bluebird Sings to the Lemonade Springs: Living and Writing in the West*. New York: Penguin, 1992.

Steinbeck, John. *America and Americans*. New York: Viking, 1966.

———. *The Grapes of Wrath*. New York: Penguin, 1978.

———. *In Dubious Battle*. New York: Penguin, 1979.

———. *The Log from the Sea of Cortez*. New York: Penguin, 1986.

———. *The Long Valley*. New York: Penguin, 1986.

———. *Of Mice and Men*. New York: Bantam, 1978.

———. *To a God Unknown*. New York: Penguin, 1986.

———. *Tortilla Flat*. New York: Penguin, 1986.

———. *Travels with Charley*. New York: Bantam, 1962.

———. *The Wayward Bus*. New York: Penguin, 1986.

———. *Working Days: The Journals of "The Grapes of Wrath."* Ed. Robert Demott. New York: Viking, 1989.

Tarter, Jim. "Collective Subjectivity and Postmodern Ecology." *ISLE* 2.2 (winter 1996): 65–84.

Taylor, Mark. *Altarity*. Chicago: University of Chicago Press, 1987.

Thomas, Keith. *Man and the Natural World: A History of the Modern Sensibility.* New York: Pantheon, 1983.
Thoreau, Henry D. *Walden.* Princeton: Princeton University Press, 1973.
Timmerman, John H. *John Steinbeck's Fiction: The Aesthetics of the Road Taken.* Oklahoma City: University of Oklahoma Press, 1986.
Tobias, Michael, ed. *Deep Ecology.* San Marcos, CA: Avant, 1988.
Tokar, Brian. "Marketing the Environment." *Zeta* February 1990: 16–17.
Turnbull, Elsie. *Trail between Two Wars: The Story of a Smelter City.* Victoria: E. G. Turnbull, 1980.
Turner, Fredrick Jackson. *The Frontier in American History.* New York: Holt, 1947.
Udall, Stuart. *The Quiet Crisis.* New York: Holt and Rinehart, 1963.
United States Congress. *A Legislative History of the Endangered and Threatened Species Act of 1973, as Amended.* Compiled in Committee Print, Serial No. 97–6, 97th Cong., 2d sess., February 1982.
Varela, Francisco. *Principles of Biological Autonomy.* New York: Elsevier North Holland, 1979.
Venturi, Robert, Denise Scott Brown, and Steven Izenour. *Learning from Las Vegas: The Forgotten Symbolism of Architectural Form.* Cambridge: MIT Press, 1977.
von Foerster, Heinz. *Observing Systems.* Seaside: Intersystems, 1981.
Wade, Nicholas. "Software for the Brain." Review of *Toward an Understanding of Consciousness,* by Daniel C. Dennett, and *The Conscious Mind,* by David J. Chalmers. *New York Times Book Review* December 29, 1996: 22.
Walton, John. *Western Times and Water Wars: State, Culture, and Rebellion in California.* Berkeley and Los Angeles: University of California Press, 1992.
Warren, Karen, and Jim Cheney. "Ecological Feminism and Ecosystem Ecology." *Hypatia* 6.1 (1991): 178–97.
Webster, Noah. "Dissertation on the Supposed Change of Temperature in Modern Winters." *A Collection of Papers on Political, Literary, and Moral Subjects.* New York: Webster and Clark, 1843.
Wehmhoefer, Richard A. "Water Law in the Southwest." *Water and the Future of the Southwest.* Albuquerque: University of New Mexico Press, 1989.
West, Robin. *Narrative, Authority, and Law.* Ann Arbor: University of Michigan Press, 1993.
Westbrook, Max, and James H. Maguire. *A Literary History of the American West.* Fort Worth: Texas Christian University Press, 1987.
Wild, Peter. "Edward Abbey: The Middle Class Maverick." *New Mexico Humanities Review* 6.2 (summer 1983): 15–23.

Williams, Raymond. *The Country and the City.* New York: Oxford University Press, 1973.
Williams, William Carlos. *In the American Grain.* New York: New Directions, 1956.
Wilson, Alexander. *The Culture of Nature: North American Landscape from Disney to the Exxon Valdez.* Cambridge: Blackwell, 1992.
Wolfe, Cary. "In Search of Post-humanist Theory." *Cultural Critique* 30 (spring 1995): 33–70.
Work, James C. "The Moral in Austin's *The Land of Little Rain.*" *Women and Western American Literature.* Ed. Helen Winter Stauffer and Susan J. Rosowski. Troy: Whitson, 1982.
Worster, Donald. *Dust Bowl: The Southern Plains in the 1930s.* New York: Oxford University Press, 1979.
———. *Nature's Economy: A History of Ecological Ideas.* Cambridge: Cambridge University Press, 1977.
———. *Rivers of Empire: Water, Aridity, and the Growth of the American West.* New York: Pantheon, 1985.
———, ed. *The Ends of the Earth: Perspectives on Modern Environmental History.* New York: Cambridge University Press, 1988.
Wright, Will. *Wild Knowledge: Science, Language, and Social Life in a Fragile Environment.* Minneapolis: University of Minnesota Press, 1992.
Wriston, Walter B. *The Twilight of Sovereignty: How the Information Revolution Is Transforming Our World.* New York: Scribner, 1992.

Index